How Can a Rational God Allow Irrational Numbers?

How Can a Rational God Allow Irrational Numbers?

An Intellectual's Guide to Faith and Christianity

JOSEPH M. WOLFE

WIPF & STOCK · Eugene, Oregon

HOW CAN A RATIONAL GOD ALLOW IRRATIONAL NUMBERS?
An Intellectual's Guide to Faith and Christianity

Copyright © 2025 Joseph M. Wolfe. All rights reserved. Except for brief quotations in critical publications or reviews, no part of this book may be reproduced in any manner without prior written permission from the publisher. Write: Permissions, Wipf and Stock Publishers, 199 W. 8th Ave., Suite 3, Eugene, OR 97401.

Wipf & Stock
An Imprint of Wipf and Stock Publishers
199 W. 8th Ave., Suite 3
Eugene, OR 97401

www.wipfandstock.com

PAPERBACK ISBN: 979-8-3852-4389-1
HARDCOVER ISBN: 979-8-3852-4390-7
EBOOK ISBN: 979-8-3852-4391-4

04/17/25

Unless otherwise specified, Scripture quotations are from the Christian Standard Bible, copyright © 2017 by Holman Bible Publishers. Used by permission. All rights reserved.

To my son, Jairus

My prayer for you is in line with David's prayer for Solomon found in 1 Chronicles 29:10–20:

> Give to my son Jairus a perfect heart to keep your commandments, your testimonies and your statutes, and to do them all, and to build on the foundation, for which I have made provision.

There is no greater aspiration a father can have than to leave a foundation of faith suitable for his children to build upon. I would like to echo Paul's words, "Imitate me, as I also imitate Christ" (1 Corinthians 11:1); however, I know my own shortcomings. I pray that you drink deeply from the Spirit, hear and heed God's call on your life, and grow into the likeness of Christ.

Contents

Acknowledgments		ix
Author's Note		xi
Chapter i	Introduction	1
Chapter $e^{i\pi}$	Identity	9
Chapter 0	(0, 0, 0, 0): Origins	20
Chapter i^i	n+1: Ode to Faith	41
Chapter 1	**N**: Supernatural	49
Chapter $\sqrt{2}$	$\partial y/\partial x$: The Critical Point	56
Chapter φ	y and y_0: Problem-Solving	62
Chapter e	$de^x/dx=e^x$: Transcendent Purpose	70
Chapter π	y: The Mystery	76
Chapter $2e$	d/dx: Spiritual Growth	83
Chapter 2π	$\int x/dx$: Spiritual Maturity	88
Chapter $3e$	A^T: The Transformation	95
Chapter 3π	x: The Variable	101
Chapter 42	ε: Humility	107
Chapter 7^3	[Set Theory]: The Church	117
Chapter ∞	QED: Live What You Believe	125
Bibliography		131
Scripture Index		135

Acknowledgments

My sincerest appreciation goes to the following people who have collaborated and participated in the development of the contents of this book through education, encouragement, proofreading, friendship, fellowship, and discipleship.

Dr. Deron Biles
    A mentor, a friend, an advisor

Dr. Eric Costanzo
    A fellow philosopher, a friend, a confidant

Dr. Barry Creamer
    A teacher, a polymath, a voice of reason

Dr. Dennis Hester
    A pastor, a friend, a counselor

Dr. James Wolfe
    A father, a foundation, a role model

Dr. Tony Wolfe
    A brother, an inspiration, a fellow student

Acknowledgments

THANK YOU

I would like to express my overwhelming gratitude to my other family members who have challenged me, coached me, advised me, and above all, despite great difficulty, always tried to understand me.

Lynelle Wolfe
>My partner in life, my love, a lifelong learner
>May the realm of Wolfetopia ever flourish.

Kathy Wolfe
>A mother, an example of unshakable faith, a survivor
>May your blind faith continue to speak for generations to come.

Terry Wolfe
>A brother, a fighter, a contender
>Remember that God has a purpose for you yet. I pray this quirky book will help you on your journey to finding that purpose.

Adam Wolfe
>A brother, an advisor, a coach
>Your challenge about merely being a pawn on God's chessboard back in our college days has stuck with me, and it has formed an integral part of my faith journey.

Jimmy Wolfe
>A brother, a believer, an encourager, a questioner
>I don't know anyone who works harder or is more thirsty for knowledge gained through personal experience. May God continue to work in you and use you through your hands-on approach to life.

Author's Note

THE TITLE OF THIS book stems from a popular philosophical question posed initially by Ivan Fyodorovitch Karamazov in Fyodor Dostoyevsky's famous novel *The Brothers Karamazov*. Ivan wrestles through a monologue explaining why he cannot come to terms with a loving God who allows innocent children to suffer. Exasperated, Ivan proclaims, "With my pitiful, earthly, Euclidean understanding, all I know is that there is suffering."[1] Ivan longs for swift justice—for immediate retribution to right the wrongs that he has witnessed. Since he cannot have this, he reasons, "If all must suffer to pay for the eternal harmony, what have children to do with it, tell me, please? It's beyond all comprehension why they should suffer, and why they should pay for the harmony."[2]

How can a loving God allow innocent children to suffer? What a profound question. Many philosophers have spent countless hours and written countless pages to explore this topic. Two such books the author would recommend are: *The Problem of Pain* by C. S. Lewis and *Can God Be Trusted?* by John G. Stackhouse Jr.

The author of the present work has attempted to explore a similar paradox—not attempting to present a new Euclidean proof of some heretofore unsolved mathematical postulate, but rather to provide a proof for the rationality of faith in the God of the Bible.

How can a rational God allow irrational numbers?

---

1. Dostoyevsky, *Brothers Karamazov*, 305.
2. Dostoyevsky, *Brothers Karamazov*, 306.

Chapter *i*

Introduction

THIS IS THE FIRST chapter of a book that attempts to logically discuss something that the intellectual community has deemed illogical. Therefore, what better chapter number for it than *i*, the imaginary number?

David Hume has wielded an enormous amount of influence over a segment of our present-day intellectual community. His philosophies regarding experience and causality have led us to reject drawing conclusions based on consequence. Hume posited that proof by induction is not pure science; it is what he called "belief." Hume's philosophy has been used to dismiss faith in (induction of) a God whom we cannot perceive or experience because induction from consequences cannot be used to prove causality (the pairing of two events only yields correlation; it does not prove causation). Hume also used these philosophical tools to refute the cosmological argument for the existence of God, contending that the world, though a functioning system, could be the result of grand-scale chance.[1]

His name has become synonymous with the humanist movement relating all science to human reason, which is the servant of human passions. Hume extended his skepticism even to doubt if

---

1. Cranston et al., "David Hume."

everything must have a cause.[2] However, this philosophy would not allow us to draw any conclusion on how the origin of the universe or life itself began. Since we cannot observe the events, we must rely on induction (belief) in describing the big bang. Neither could we derive causation of climate change from a correlation between it and human behavior. The two events may be related by contiguity and timing; however, to assume one event caused another requires faith. In fact, anything we may presume to know is actually a belief we hold.

The positivists or pure empiricists have influenced a second segment of our mainstream sciences and present-day intellectual community. These rely strictly on the objective sciences like mathematics, physics, and chemistry to pursue truth. According to this philosophy, all truth must be derived from empirical data (from experience) or from logical reasoning. Among the early followers of this particular movement are some curious characters. Auguste Comte started his own short-lived religion worshipping humanity. Wilhelm Ostwald, following the positivist empirical view, denied the existence of the atom because it could not be seen. Hans Vaihinger also rejected the concepts of the infinite and the infinitesimal.[3]

This faction of thinkers is faced with two primary challenges. The first is illustrated by the examples of Ostwald and Vaihinger. What defines objective observation? The science of observation and laboratory testing is repeatable; however, how the results are interpreted depends on the *subjective* experience of the scientist. The author recently had a discussion with a brilliant scientist who explained that he had performed repeated tests over a decade to demonstrate a specific phenomenon; however, certain laboratory scientists did not believe his data or his research until they had spent nearly two years repeating the experiments themselves. This delay in accepting data and research and the interpretation thereof, which cost the associated industry valuable time and money, was a denial because of lack of personal experience, not because

2. Hume, *Treatise of Human Nature*, 78–82.
3. Feigl et al., "Positivism."

INTRODUCTION

there was no empirical evidence. The second challenge is the subjective interpretation of statistical data and its incorporation into predictive models. Simulation models provide nonunique solutions because they entail a number of variables, many of which are codependent. In statistical analysis, these dependencies between variables are not always identified, depending once again on the subjective experience of the scientist. The method used to history-match these statistical data depends on how these variables are treated within the model, and thus two modelers may weight variables differently in their interpretations and generate very similar history matches but derive very different forecasts. Furthermore, models are often built to describe small, contained environments (sector models) where material balance may be demonstrated, predicted, or controlled. How these models are then scaled up to reflect the larger reality is also a subjective process.

The post-positivist may also be influencing a third segment of our intellectual and scientific communities by informing us that our philosophers and scientists are all biased. Because this system formed in response to the critiques to positivism, these critical realists subjectively observe that a scientist's observations are subjective rather than objective.[4] This philosophical view suggests that no science is *settled*, that hypotheses cannot be positively proved by a finite number of positive experiments, nor can they be disproved by negative experiments, since human error could lead to faulty results.[5]

This line of reasoning, however, may also lead us to reject the basic mathematical principles as subject to human observation. Are all facts subject to human observation? Can we definitively say that 2+2=4? Should we ban the teaching of arithmetic because it subjectively favors one group of cultural thinkers over another? This particular branch of reasoning has a tendency to be politically influenced more by the social sciences (psychology and sociology) than by the scientific method.

4. Mann, "Post Positivism."
5. Difference Between, "Positivism and Post-Positivism."

How Can a Rational God Allow Irrational Numbers?

These conflicting philosophical views form substantial bases for scientific study. The author hopes that the reader, having considered these philosophical views, has developed a healthy dose of skepticism for what our mainstream scientific community teaches. In one regard, the Humean philosophy concludes that everything we can know is rooted in belief. The positivist is left to trust only science that he/she alone has been able to prove through personal experimentation, because one cannot trust other scientists, as they may be charlatans, unless we perhaps build our own religion and convince others to trust our observations. And finally, the postpositivist suggests that all truth is relative to the observer. This is quite a conflict. It should, in fact, leave the reader with a true crisis of faith, though we don't want to call it by that name.

The challenge we as intellectuals face in accepting faith as a sufficient source of proof is our unwillingness to concede that not everything can be explained by physical laws or clearly delineated by human reasoning. In fact, we have gone so far to the other extreme in our confidence in human knowledge or observation that we reject or dismiss plausible explanations because they defy our preconceived notions, much as a scientist may force data to validate an initial theory by reinterpretation or over normalization (confirmation bias). At first this process is subtle, but as the data continue to be collected, the legitimate normalization and transformation of these data become indefensible; they have become a snowball, and funding may be dependent on results (publication bias). Looking back, the scientific methods we hold near and dear begin to resemble data manipulation. The inconsistencies begin innocently enough through unintended data dredging and survivorship bias, but they eventually evolve into irresponsible gerrymandering. Who will hold the mainstream scientific community responsible?

As an example, the concept of the big bang contradicts the established physical laws of entropy. Interestingly, if we are willing to accept the big bang, which necessarily occurred to form the universe and which defies natural law, why then are we unwilling to call it a *super*natural force? This is sleight of hand or smoke

Introduction

and mirrors. The mainstream scientific community ignores the inconsistency of order and chaos and misleads the rest of us. What happened before the big bang? Where did the high-density material that exploded originate? Suppose it spontaneously formed from the Nothingness. Then it necessarily formed *super*naturally, because nothing comes from the Nothingness, according to the laws of conservation of mass and energy. Suppose on the other hand that matter is eternal. Then how can we dogmatically ascribe an age to the universe?

The mainstream scientific community has disingenuously led the intellectual community to trust in human understanding. This may be driven by a desire for money, power, or fame, perhaps to comply with some political agenda, or maybe just to stroke one's own ego. Whatever the motive, the result is the same. The mainstream scientific community has been irresponsible to humanity, having departed from the high purpose of science to seek general truth from the study of the physical world. Once the physical world exhibits behaviors contrary to physical laws, it is prudent to continue study to determine if unknown laws are at work. However, we are beyond that stage; we are now to the point where, to borrow a cliché, we are assuming there can be no black swans because we have never observed one (negativity bias). In fact, the author has personally seen a black swan. We have determined that our preconceived understanding of the physical laws and our knowledge must be sufficient to explain the universe, rather than relying on our understanding and knowledge to reach the limits of our ability and honestly admitting that we cannot explain certain events or data, such as the origin of the universe.

Think of this book as representing the universe of knowable facts and information. Now, consider that the knowledge that we actually possess as humans fits exhaustively in the dot over this letter *i*. O the depths of the wisdom and knowledge of humanity concentrated in the tittle! From this infinitesimal amount of knowledge, we extrapolate all the words and blank space of this book. Imagine trying to use the tittle, one single data point, in some word-statistical and -spatial analysis. In geostatistical analysis, we

How Can a Rational God Allow Irrational Numbers?

could use univariate kriging and multivariate cokriging to improve our interpretation;[6] however, we would need several data points to rely on kriging (such as other letters distributed on each page), and at least some of those data points would require a second type of data (such as lengths of words) to employ cokriging. From only one data point, the best we could hope to draw is a target, a bull's-eye. That is exactly what happens with human intellect: we collapse into a self-centered black hole. Imagine that by some miracle, we were able to extrapolate the contents of this book from the single tittle; even so, we have managed only to define the knowable. Consider the room you are sitting in as the universe of all knowledge.

There are four types of knowledge: there's the (1) known known, (2) the unknown known, (3) the known unknown, and (4) the unknown unknown. The first group contains facts that we consciously learn, such as arithmetic, the principles of science, and recorded historical events. The second group consists of our idiosyncrasies, muscle memory, the unconscious knowledge we hold. In the third group are the principles that are within our grasp to be derived or experienced; those that we are conscious of but have not identified. The fourth group, however, represents the things that we don't even know exist; we have no inclination of this void in our knowledge. Imagine extrapolating a sine function from the data points of only a half period. Whatever the latest trend we witness will be errantly projected forward. Now imagine that not only are we errantly extrapolating our two-dimensional knowledge into the content of the book, but there are multiple dimensions that we are not even aware of within the room.

In other words, it is primarily our ego that keeps us from logically concluding that, based on all available data and information, something *super*natural occurred at the beginning of the universe. If we could get to that point, it would be possible to accept that other *super*natural events such as the virgin birth and the resurrection from the dead might have occurred.

In fact, *science* has become a sort of religion requiring faith in its favored priests and prophets, who command and disseminate

---

6. Chambers and Yarus, "Geostatistical Reservoir Modeling," 68–73.

INTRODUCTION

the tenets of their doctrine in the form of papers, museum exhibits, school curriculums, videos, and political initiatives. The terminology used in this religion is attractive to intellectuals because it appeals to our egos; it sounds intelligent. One who regurgitates the mantra with other *believers* is received readily as a scientific savant. The continual propagation of this doctrine serves to reinforce and validate itself; it is incestuous. This is an example of what Jesus calls "the blind guiding the blind" (Luke 6:39). We have, in fact, fulfilled Paul's description of us from nearly two thousand years ago: "Professing themselves to be wise, they became fools" (Romans 1:22).

Our current situation is not unlike what Ignaz Philipp Semmelweis faced in the 1800s. The mainstream scientific community of the day (including his medical superior, as well as professors and the editor of a prominent medical journal) disregarded his observations because they could not see and explain them. They rationalized and used their understanding of physical law and their overwhelming knowledge to dismiss Semmelweis's observations and ridicule him.[7] How many mothers died in childbirth because the doctors of the day could not see germs? These women had *faith* in the scientific knowledge and understanding available at the time. While the scientific method eventually won and *corrected itself*, this does not change anything for the women who died; they died in their faith. In other words, in our lifetime, science will evolve; and whatever we trust and believe today, based on scientific interpretation, may change after we are gone. Faith is not a new concept for the intellectual; it parades itself proudly in superiority, merely masquerading as intelligence. However, it is misplaced in human knowledge.

If it were not evident previously, it should be by now that as an intellectual, you have been placing your faith, perhaps unbeknownst to you, in the mainstream scientific community. As I have unabashedly explained, the mainstream scientific community is undeserving. Since you are now aware that you, as an intellectual, have been willingly practicing faith, I hope you will

---

7. Zoltán et al., "Ignaz Semmelweis."

be an open-minded, free thinker as you read the rest of the content in this book.

The next chapter, chapter $e^{i\pi}$, addresses identity and worldview, directing the reader to consider how beliefs are correlated to and causative of behaviors. The following chapter, chapter o, addresses the origin. It is a general explanation of the uncertainties surrounding the origin of the universe. This discussion is intentionally a philosophical view of the scientific interpretations and the conclusions of the same. It is not intended to be an exhaustive dissertation on the subject matter but rather a practical analysis and logical reasoning of the ramifications. After this, the book progresses to explain what faith is and why it is a reasonable response to unexplainable events and then to describe some tenets of the Christian faith.

It is not the author's intent to present a proof for the existence of God. There are several scholarly, philosophical arguments on both sides of the subject; and it is the author's belief that such a proof is beyond the capabilities of human reasoning. It is also not the author's intent to demonstrate or prove the veracity of the Bible. There are various historical, archaeological, and scientific studies and a significant amount of evidence available with which the author is satisfied. The scope of the present work is to present and expound on the particulars of faith and Christianity into the intellectual vernacular.

Chapter $e^{i\pi}$

Identity

Euler's identity, $e^{i\pi}+1=0$, is one of the most incredible equations in all of mathematics. Its elegance is complemented by the inclusion of arguably the five most important numbers: 0, 1, e, π, and i. Complex analysis, involving imaginary and real numbers, may seem to be the most insignificant of mathematical branches since there appears to be little use of imaginary numbers in the real world; but what we see through this equation is that the imaginary number is transported into the real world. It is because of Euler's identity that i^i is a real number and perhaps that *imaginary* faith (chs. i and i^i) can, in fact, become sight.

Identities play a key role in mathematics. For instance, 0 is the additive identity, since $a+0=a$; and 1 is the multiplicative identity, since $a*1=a$. There are various types of mathematical identities found in geometry, trigonometry, calculus, etc. But what is an identity? In mathematics, an identity like Euler's identity or the trigonometric identities, such as the law of sines and the law of cosines, are statements of definition or of truth; this type of identity represents an axiom to be used to demonstrate or derive other truisms.

Shakespeare, through Juliet, ventured to ask and proclaim, "What's in a name? That which we call a rose by any other name

How Can a Rational God Allow Irrational Numbers?

would smell as sweet."[1] Here Shakespeare entertains a philosophical question as to whether one's identity is defined by his/her name. We may extend this probe to inquire, "Are we defined by the labels we choose for ourselves or those others choose for us?" Over the past decade, the author has been amused by the irony of our American society as various groups of individuals have proclaimed outrage over being stereotyped by others and then simultaneously labeling (stereotyping) themselves with icons and monikers on their social media platforms and even dressing accessories. When one elects to outwardly align themselves with a cause, it is only natural for onlookers to associate identity with these labels. And, when the observable facts (such as anatomy or behavior) do not corroborate the identity claim, it is logical that onlookers would question the veracity of this truth claim. To ignore these inconsistencies would be delusional or hypocritical at the very least and possibly abusive and manipulative.

René Descartes is considered to be the father of analytical geometry and the father of modern philosophy. When he considered the question of identity, he penned the famous line "Cogito, ergo sum" (I think, therefore I am) as part of his meditations on philosophy.[2] This proclamation of self was a well-reasoned argument based on empirical observation. Because I (whatever "I" entails) have the ability to cognitively process and even ponder my own existence (think), I must in fact exist (as anything that has the ability to think must necessarily exist). Descartes was a rational man who spent time questioning even his own conclusions. His basis for reasoning and drawing conclusions, however, was factual observation or natural philosophy.

In society today, we have lost this ability to reason and question. We have lost the objective art of observation and sacrificed philosophy for psychology. Our culture has twisted Descartes's words from the philosophical, observable, objective truth of existence into a predication of psychological, unobservable, preferred state of being. Rather than "I think, therefore I exist," a rational

---

1. Shakespeare, *Romeo and Juliet*, 2.2.71–73.
2. Watson et al., "René Descartes."

and observable conclusion, our culture abuses the philosophical argument to state things such as "I think I am a woman, therefore I am a woman," an unverifiable, desired status. The reader may fill in the blank as he/she chooses. In other words, rather than relying on the faculties of reason to evaluate the reality of our observable existence, our culture has determined to use our reasoning to define alternatives to our observable reality.

It is common for humans to aspire to accomplish or become more than they perceive themselves to be. Such desire spawns creativity and innovation. For instance, the Wright brothers did not decide psychologically that they were birds in order to fly. No, their aspirations led to countless hours of research and experimentation. Realization of their dream of flight depended precisely on their ability to work out the physics of aeronautics, not in identifying as large flying mammals.

So, how should one define his/her identity? Many prominent philosophers other than Descartes have also addressed this question. Aristotle would have us believe that the human self consists of an inseparable body-and-soul unit. While the soul is the essence of the "true self," enabling us to live, sense, reason, and function, the body represents the matter of the soul.[3] Descartes suggested that we exist as a dual entity—that mind and matter are distinct. The mind (soul), he concludes, is immortal because it cannot be divided into parts and exists independent of the body, while the matter (body) can be broken into smaller parts.[4] John Locke posits that self is defined by the ability to reason, reflect, and conceive of itself as an independent entity. Personal identity is "the sameness of a rational being" or the continuity over time of the entity.[5] David Hume, in contradiction to Locke, concludes that there is no consistency in our ideas and imaginations to define the entity of self. "The constancy of our impressions, i.e. their resemblance at different times, makes us consider them individually the same . . . a succession of related impressions places the mind in the same

---

3. PHILO-Notes, "Aristotle's Concept of Self."
4. Watson et al., "René Descartes."
5. Piccirillo, "Lockean Memory Theory," para. 2.

disposition as does an identical object . . . , and so we confound succession with identity."[6]

The distinct viewpoints on how we define *self* are interesting, especially while considering questions of our physical and rational identities. However, while these philosophical arguments over identity and *self* are of significant import, they are not the core issue in our present discussion. In fact, the key questions for our consideration concern the basis for our reasoning, our worldview. How do we see the world, and what is our relationship to that world? Of course, those in and of themselves have been questions for these same philosophers and many others.

Let us return momentarily to the popular tool of psychology and to sociology. One's behavior depends heavily on one's perception of reality. Some of these aspects are referenced in the text of this work, such as students' unwillingness to study material when they believe it will not be useful in the future (ch. 2e) or the author's shopping habits while being observed (ch. 3π). Consider, however, how cultural influences affect the way individuals observe events, draw conclusions, and behave.

In her article "How Culture Affects the Way We Think," Catherine West cites various authors whose research demonstrate the differences in brain activity associated with cultural norms, whether that be between peoples from Eastern and Western cultures or even within different segments of the same culture. West states that research findings by Qi Wang of Cornell University "illustrate that constructs of the self differ across cultures as a function of the social orientations, cultural values, and narrative environments in which children are raised."[7] There is a significant body of work on cross-cultural psychology interrogating topics such as individualism versus collectivism and personalities within different cultures.[8]

By observing these types of differences in adult populations and associating them with the cultural components that led to a

---

6. Hume, *Treatise of Human Nature*, 666.
7. West, "How Culture Affects," para. 15.
8. Cherry, "What Is Cross-Cultural Psychology?"

Identity

desired end, it would be possible for a small group of people in positions of political or military power to manufacture a future culture by carefully directing the education and beliefs of the current society's children. A government would need only to (1) control the educational curriculum of all children from three to eighteen years of age, (2) convince parents that the quality of that education would benefit their children as well as society, and (3) remove the parents from an active role in questioning the fundamentals of that educational system. How easy would it be to manipulate these children into a society of obedient slaves within twenty to thirty years? Interesting.

Our environment and education certainly play a role in how we define ourselves; the age-old question of nature versus nurture could be raised. But how should one identify himself/herself simply? Let's ask John Smith the question "Who are you?" John Smith may respond in any number of ways, such as:

- I'm a man.
- I'm a plumber.
- I'm a Smith like my father before me.
- I'm an American.
- I'm a Democrat (Republican, Independent).
- I'm a husband and father.
- I'm an atheist (agnostic, Buddhist, Muslim, Christian).
- I'm just some random guy (the author's preferred response).

Does any of these responses really answer the question? Is a person the sum of certain attributes of his/her life or characteristics of his/her behavior? Recently, the author met a friend's girlfriend for the first time. The author asked her the question "What do you do?" She responded by identifying several of her hobbies. An observer later commented, "She never did answer your question; maybe she didn't understand that you wanted to know what her job was." The author responded, "I didn't really care how she *answered* the question; it was intentionally ambiguous because I wanted to see how she would *interpret* the question."

It is the author's opinion that how we interpret the questions "Who are you?" and "What do you do?" is a more indicative response than the particulars of our answers. Of course, how we interpret a question posed to us is dependent not only on our cultural and social experiences but also on our personalities, our situational awareness, the circumstances or ambiance in which the question is posed, and who we believe is asking the question. This is precisely why the author comes to his conclusion that how we interpret the question tells more about us than the words we formulate in response. (The reader may appropriately conclude that the author overanalyzes every situation and probably wears himself out with all of these mental acrobatics. Perhaps this is the author's identity.)

While Euler's identity is a truism that bridges two realms of number systems, the imaginary and real numbers, our identities are truisms that bridge our beliefs and our behaviors. When the imaginary and real numbers appear together, such as $2i+7$, we call this a complex number. Complex analysis involves working equations with complex numbers. Most of our lives are lived in complex situations where we attempt to balance our beliefs and our behaviors. As you will read in chapter 2π, the author holds a simple view on maturity, where one's values (beliefs) align with one's actions (behaviors). We are not defined or identified with our ideological professions of social justice or self-sufficiency. We are defined and identified by the confluence of our values and our actions. When these two functional systems align consistently, we have achieved maturity. While these two systems are at odds with one another, we are complex beings negotiating between who we desire to be and our present state of being, which doesn't always please us. As we mature, our beliefs and behaviors may change in order to bring balance (ch. $3e$) to our lives and to achieve maturity.

The additive and multiplicative identities (0 and 1, respectively) are not truisms like Euler's identity or trigonometric identities. The additive and multiplicative identities are special values that allow the addend or factor to remain themselves in spite of the operation applied to them. Remember that $a+0=a$. The additive identity 0 leaves a unchanged by the addition operator. Likewise, the multiplicative identity 1 leaves a unchanged by the multiplication operator. This distinct definition of identity is interesting because these identities allow one to remain constant and unfazed by circumstances or outside forces. Finding one's identity should be liberating; it should allow him/her to be who he/she is. However, in our present culture, our children are encouraged to be disgruntled with who they are. We have individuals who deny their ancestry, heritage, skin color, gender, and even their own previous statements. All of these can be seen in prominent circles such as politics, entertainment, and academics. What are we teaching our children?

During the author's childhood, there was a popular cultural slogan inspiring young people to "Be what you want to be." The irony of the slogan was not lost on the author, since culture would then explicitly follow this statement with a precise depiction of what it was that "you want to be." According to our culture at the time, you wanted to be rich, tall, skinny, sexy, big busted, famous, etc. Now, this slogan has expanded even beyond the human realm, and culture is still pushing the boundaries as to "what" you want to be. It is the author's opinion that this is a result of the bases of our cultural worldview. Because beliefs affect behaviors, the philosophical foundation that society lays for our children will determine how stable the future society will be. Philosophy forms beliefs, which control psychology, which affects behavior. If our society does not present a consistent set of core values of what defines humanity and the world, our children cannot learn to function within the appropriate boundaries.

Several years ago, while watching a nationally televised Major League Baseball game on television, the author witnessed disturbing commentary by a guest TV announcer. A batter was at the plate, the pitcher released the ball, and the ball hit the knob of the bat below the hands of the batter as he attempted to dodge the ball. The batter realized what had happened, as did the umpire, who

How Can a Rational God Allow Irrational Numbers?

called the ball foul, and the game moved on with one more strike in the count. The guest announcer in question, a former player himself, proceeded to criticize the player for not faking an injury. "He should have grabbed his hand and pinched his finger to make it look like the ball had hit him instead of the bat. In this game, you're just trying to win, and you do whatever it takes to get on base and help your team win."

Earlier in the season, a no-hitter was spoiled when a batter feigned having been hit by a pitch. The author witnessed another player take first base under a similar guise on a less monumental occasion that same season. Are these not questions of character and integrity? Do these not reflect on our society as a whole? What lessons are we teaching, and how far do these lessons extend?

Is a used car salesman not merely trying to help his company and family thrive or survive by lying to a customer? Why is it okay to "mislead" an umpire and an entire stadium of fans, all to the potential demise of the opponent's opportunity to succeed? "Oh, it's just a game and nothing depends on it," the reader may respond. Remember, however, that professional athletes are paid exorbitant salaries and are (whether we like it or not) role models for our children. If they are to be referred to as professionals, we should expect them to act professionally. The batter in the televised event demonstrated his integrity by being honest and not feigning injury. If professional athletes are not held accountable for their actions, why then was Bernard Madoff sent to prison; wasn't he just trying to help his home team win? How do our children navigate this duplicitous system of accountabilities?

The next chapter (ch. 0) explains the ramifications of believing that our lives are the result of cosmic chaos and random chance. If that is our philosophical foundation for the origins of humanity, then why do we have sociological norms at all? Who is to say your son cannot be a domesticated cat instead of a boy? Why is cannibalism frowned upon? What makes indecent exposure indecent? Why is slavery wrong? Why should we care about climate change? Why are Ponzi schemes bad? Are the responses to these questions mandated by society-defined morality?

Identity

Imagine if we attempted to impose the rules of chess on a game of checkers. Even though these games share the same playing board, they have different pieces, tactics, and objectives. Now imagine governing a game of mancala by chess rules. These two games don't even share the same board. Society-defined morality suggests that whatever a society deems to be moral at a given time is moral. Whatever a society believes to be good (subjectively, as there is no objective measure) is good.

If a society defines morality, then if there is some future society that esteems cannibalism to be good, then cannibalism is good in that society, as their morals are defined by their society. Furthermore, if a current society in a foreign place determines that cannibalism is good, then cannibalism is good in that society. One step further, if a previous society of Americans had determined that cannibalism were good, then cannibalism was good at that time until American society changed its opinion.

Much like governing the checkers and mancala games by chess rules, as long as we subscribe to the society-defined morality mantra, one society cannot judge another society. Americans cannot judge future or past American societies or any foreign societies based on our current morality, because each society's morality is completely defined by its own society.

In other words, to subscribe to society-defined morality, we must agree that slavery was good while society approved of it, and homosexuality was wrong until society approved of it. Americans should not meddle in the affairs of other nations (including in the world wars), nor should we attempt to influence other nations. When Martin Luther King Jr. stated, "Injustice anywhere is a threat to justice everywhere,"[9] he had no basis for this statement, since there is no ultimate authority to define injustice and justice. To disagree and say that a former society was wrong about their assessment is to trade their oppressive set of norms for your own, as well as to completely reject the originally professed position that society defines morality.

---

9. Letter from Birmingham, Alabama jail, Apr. 16, 1963; quoted in National Park Service, "Martin Luther King, Jr. Memorial," s.vv. "North Wall."

How Can a Rational God Allow Irrational Numbers?

Society-defined morality is the epitome of groupthink by its very definition. Let the intellectuals cringe at their colleagues who profess mindless submission to so-called society-defined morality. True, rules of conduct must be established by someone. If there is no outside entity to set objective standards, then we are indeed all to be governed by this mindless groupthink where we can rise to the level of our collective incompetence. This response is akin to one's parental unit responding to the question of why one should clean his/her room with the unsatisfactory response "Because I said so."

In recent years, American society has grieved as we have witnessed many suicides in the celebrity world as well as in our individual families. This seems to be a growing trend, according to the Centers for Disease Control, which cite a 36-percent increase in suicides from 2000 to 2022.[10] Depression and mental illness are two of the leading causes of suicide. The author acknowledges the gravity of loss of life to suicide and the hopelessness that weighs on an individual who contemplates suicide. However, as a society, we should question if the undesired behaviors we are witnessing are a response to a cultural belief that we have been proliferating.

In a *because-I-said-so* culture, our identities are defined by whomever is allowed to answer with this dreaded response. This is the opposite of a liberating identity, as it represents submission and subjection to some unexplained reasoning; and it is certainly not an axiom to enable derivation of other truisms, because future direction will be based on whimsical, subjective decisions by some unpredictable, external entity. Under this totalitarian regime, we cannot repeat Descartes's philosophical argument, because we are forced to surrender our cognitive abilities in favor of groupthink. Nor can we subscribe to Aristotle's or Locke's philosophies because of our lack of independent reasoning. We are left only with Hume's depressing conclusion that there is no *self*, no consistent form of reasoning ideas that we could define as an individual's essence. There is only existence, no being. In a culture that actively teaches

---

10. CDC Suicide Prevention, "Facts About Suicide."

this philosophy, why should we be surprised or appalled by an increase in suicides?

Our culture is apparently immature according to the author's simple definition because it does not desire the psychology and behaviors of the philosophy and beliefs it holds. This is not a new revelation; Richard Weaver explained this in his book *Ideas Have Consequences*. To determine who we are, we must know our origin. We must define our philosophy clearly and determine whether the resulting behaviors are consistent with our core beliefs.

Chapter 0

(0, 0, 0, 0): Origins

NOTHINGNESS, THE ABSENCE OF matter and energy. In the Nothingness, only logical, mathematical principles exist; only logic exists, independent of the mind to comprehend it. Spontaneously, a physical object appears out of the Nothingness, the simplest of atoms—hydrogen with one proton and one electron. This is the first singularity event, where matter and energy formed from the Nothingness. Another atom appears from the Nothingness. It is at this point that physics is born, as it requires the existence of matter and/or energy to interact. Physical changes begin to bring about chemicals, which then undergo chemical changes, resulting in the birth of chemistry. At this point, the atoms and chemicals are all compressed in an impossibly small, finite space, exponentially growing and generating energy in a tiny, high-density particle. Eventually, this high-density particle cannot sustain its compaction or construction, so it explodes with a big bang into an infinite space, ever expanding.

The term *logic* is used here to describe the apathetic construct that governs the natural behavior of the universe. This logic must necessarily exist to govern nature even in the Nothingness; otherwise a conundrum exists in that the governing force, whatever one chooses to call it, must develop itself, which is circular. These include the fundamental principles of cause and effect

(0, 0, 0, 0): Origins

and mathematical axioms. The potential for the supernatural is dismissed from consideration in the premise of this argument, though, were the supernatural being to exist, there is no obligation for logic to preexist the supernatural being, as it could be resultant from such.

The origins as stated defy the physical laws of the conservation of mass and the conservation of energy, which state respectively that neither matter nor energy can be created or destroyed. Matter and energy can be transferred from one form to another, but neither can be naturally created from or obliterated to the Nothingness. Therefore, these origins are necessarily supernatural, which contradicts the premise. This assessment also ignores the argument of the eternal existence of matter and/or energy. This position, presented by philosophers as early as Aristotle,[1] resolves the messy supernatural forces required to overcome the physical laws in some respects, but it is a complex explanation as well, since one must now reconcile the origin and age of the universe. Specifically, when discussing the age of the universe (see the section on time), if matter is eternal, then in fact, the universe is ageless, which leads us to question how there is energy left at this moment in time, as it should have been transformed into lower and lower levels, eventually dissipating. Some philosophers have argued that matter is absolutely eternal; others have argued that matter is absolutely not eternal; and still others have argued that we cannot know whether matter is eternal or not, and either position requires faith. Assertion of the eternal existence of matter or energy, while posing its own conflicts, does not negate the argument presented herein. The origin denoted here as the Nothingness could begin with the Lifelessness.

In the expanse of space, all of the matter and energy are dispersed. Then blind forces bring together the necessary gaseous chemicals in the proper proportions, and an electrostatic charge excites them, resulting in complex amino acids to form proteins, which are required to bring about the birth of a living, single-cell organism and, along with it, biology.

1. Ainsworth, "Form vs. Matter."

How Can a Rational God Allow Irrational Numbers?

This description of the origin of life is credited to Alexander Oparin and J. B. S. Haldane, who independently presented the hypothesis in the 1920s. It was not until the 1950s, however, that Stanley Miller and Harold Urey tested the hypothesis through a specifically designed experiment.[2] If these chemicals were to exist by process of these origins, and then were to react in order to produce the necessary amino acids and proteins, the odds of gathering them in the proper proportions in an instantaneously secluded space in the infinite reaches of the newly encountered universe (in our case, in an ocean on Earth) where lightning could strike without an intentional design are beyond astronomical. The absence of an intelligent mind to orchestrate this event means that elements must necessarily randomly combine repeatedly, without learning from previous errant combinations. This argument concerning the probability assessment of the given event is generally presented and dismissed from an end perspective; that is, working backward, we consider what is the probability that the initial state randomly convolved, resulting in what we observe today. However, while the odds are astronomical, we are also looking at an astronomical scale, and the initial state had no intention of resulting in what we observe today. Still, there are other credible challenges to this process generating life. As Dr. Charles McCombs notes in his May 2004 article, the referenced experiment did not conclude with life but rather in amino acids, which are necessary for life. Furthermore, there is a three-dimensional structure of chemicals that, according to him, "destroys the claim that life came from chemicals."[3]

EVOLUTION

The single-cell organism evolves into complex organisms by effecting one small change at a time. Consider how the evolutionary process would function on a computer program. Imagine the program

---

2. Biological Principles, "Origin of Life."
3. McCombs, "Problem with Chirality," para. 3.

(0, 0, 0, 0): ORIGINS

is alive, and once the program fails to compile, it dies, requiring the entire program to be randomly reconstructed. Each character of the program is one evolutionary step. Can a small change be made and the program still survive? No. Assuming that somehow the evolutionary process was able to write three lines of functional code to produce this living program, once one character of code is added, the program fails. (Ignore the very basic assumption that the origins above also had to produce an environment that could sustain, compile, and interpret the code.) This process of making one small change at a time assumes that the evolutionary process has some memory of historical success and failure so that it can not only randomly return to some functional construct in the evolutionary progress but also make more than one single change at a time. Is this a reasonable analogy to the conception of life and evolutionary process?

The arguments against the random-chance beginning to life are based on a probability assessment of the infinite number of steps that necessarily occurred in proper order to have conceived life. Dr. Henry Morris presented an article in November 2003 discussing the history and facts of these arguments.[4] However, tests of selected events have been performed to refute or at least mitigate this probabilistic argument. These computer simulations and experiments rely on ergodicity to demonstrate that evolution could, in fact, have defied these odds because of the economy of scale and natural selection's habit of allowing only certain combinations to survive and reproduce.[5] This means that instead of considering the odds of a single fair die of one million sides cast repeatedly over time until a one turns up one million consecutive times, the universe has sufficiently large space such that an infinite number of fair dice of one million sides could be cast simultaneously and repeatedly until at least one of those dice became evolution's golden ticket to life. This suffers some minor setback since science has identified the Goldilocks Zone, which refers to habitable regions

---

4. Morris, "Mathematical Impossibility of Evolution."
5. Bailey, "Do Probability Arguments Refute."

How Can a Rational God Allow Irrational Numbers?

of space with just the right conditions for life to survive.[6] While our understanding of the Goldilocks Zone is advancing and the zone itself is expanding, this concept causes a potential limit in this explanation, since the evolutionary forces would no longer have the same boundless playground for dice to be cast (a severely diminished space for those dice equates to a significantly reduced number of dice). Furthermore, the time component becomes even more critical. These events must not only have occurred in a Goldilocks Zone, they also must necessarily have occurred within a certain time frame when the elements to sustain life would have been developed within that Goldilocks Zone. However, even the introduction of the Goldilocks Zone is just another variable to be added to our probabilistic assessment. All in all, this calculation of the odds is inconclusive. If life exists, it began somewhere; with our premises, nature defied these odds; and we are generally interested in the what and how—not the why. Improbable is not the same as impossible. However, there does appear to be an inconsistency in our estimate of the age of the universe. According to a scientific paper published by the *Journal of Cosmology and Astroparticle Physics* in August 2016, "Although our result puts the probability of finding ourselves at the current cosmic time within the 0.1% level, rare events do happen."[7] Perhaps we are indeed ahead of our time? This inconsistency is somewhat troubling. If on the other hand, we subscribe to the eternality of the universe and matter, all the energy of the universe should have been transferred into useless energy long before now.

The single-cell organism evolves into complex organisms by effecting one small change at a time. This process continues until some of these become sentient beings. Does an amoeba have the ability to perceive its environment and locate tasty morsels of bacteria? Is a plant's natural response to sunlight, water, or fertile ground evidence of perception? Certainly, we would agree that at least some animals have the ability to perceive their surroundings. But where is this distinction drawn? If we conclude that the

---

6. NASA, "Goldilocks Zone."
7. Loeb et al., "Relative Likelihood for Life," 8.

(0, 0, 0, 0): Origins

single-cell amoeba is ingrained with sentience, then perhaps this capacity to perceive is synonymous with life.

At some point in this process, the sentient being passes through a new evolutionary step when it begins to respond to its surroundings. Here a second singularity of consciousness has occurred; this is a discontinuity, a huge jump in the development curve. Whether biologically or psychologically, the conscious being is able to respond to its environment and circumstances rather than merely obey a defined evolutionary instinctual system of behaviors. How did this awareness evolve? Was it a gradual process or an instantaneous one? In the past decade, a new theory, the *attention schema theory*, was introduced to suggest that the brain's neural network adapted in response to receiving increasing amounts of data (through perceptions, so is a central brain required for perception?). The receptors of the brain compete for attention to the information each has received in an attempt to trigger a response, and the brain evolved to process all of these signals appropriately. According to this theory, the process was gradual, over a half-billion years.[8]

In philosophy and even social sciences, the term *consciousness* may be used more frequently to refer to human awareness; however, the author here uses the word more loosely to denote living organisms who have the ability to observe and respond to external forces rather than functioning according to an evolutionary program. Even in the previous statement, the author is expressing bias, implying that the evolutionary process has been able to produce an organism that behaves according to a system exterior to evolution. For this author's purposes, communication is a key indicator of consciousness, as will be apparent in the next section. The discontinuity of the evolution of consciousness has received little attention in the scientific community; however, the new theory referenced suggests that consciousness is present in a range of vertebrates, which is consistent with the author's use in this document.

---

8. Graziano, "How Consciousness Evolved."

How Can a Rational God Allow Irrational Numbers?

Once this stage has been reached, evolution must necessarily be divided into two branches: unconscious and conscious.

WAR OF TWO NATURES

With this self-awareness, psychology is born. Once a group of conscious beings forms a community, sociology is born. These members of the new community alter themselves and manipulate others. Consider the occurrence of winter: Unconscious evolutionary processes could effect change over generations of species while conscious evolutionary processes could effect immediate change. Unconscious evolution will preserve only the physically fit through the process of natural selection. Conscious evolution allows for the protection of a larger portion of the population. A species of birds may evolve to winter with additional feathers over generations if the species were to stay in one location and experience the severity of weather, in which many individuals would perish. On the other hand, the same species may learn to migrate in response to their perceptions of the environment, to protect themselves and allow the weaker members of the species to continue to reproduce. With such observation, it becomes apparent that, at the very least, conscious beings can weaken their own species physically by the very nature of the mental strength evolution has bestowed upon them. Interesting.

In the animal kingdom (and by this the author means to exclude humans, though through the bases of this argument there is no such distinction), conscious manipulation of others inside the species is displayed through physical strength—as the beating of the chest by the ape or stamping of hooves by the bovines, as well as mental deception—as feigning injury to lead a predator astray by some birds or hiding from predator or prey, though the latter may be a result primarily of using unconscious physical evolution (natural camouflage) to one's own advantage by sheer accident rather than by conscious response to a threat.

The reader may accuse the author of gross animation or dramatization of the effects of psychology in the evolutionary process rather than attributing these to the natural, biological, unconscious evolutionary forces. However, these animal behaviors represent a conscious response to the entity's circumstances. The author concedes that natural camouflage might have been developed by unconscious evolution, and accidental evasion could be the result. However, it would be difficult to explain how a bird would feign injury[9] aside from psychological manipulation of the predator resulting from consciousness rather than unconsciousness.

In the human species, it is readily observed that manipulation is used in various forms unidentified in the animal kingdom—physical, psychological, mental, sexual, emotional. If the knowledge of oneself that conscious beings possess does in fact grant a species the ability to physically weaken itself as postulated, then humans are the greatest offender, as we have gone beyond physical alterations, such as relocating, layering clothing, and building shelters; we have devised chemical means to support the weaker members of our species and, beyond that, we have formed social constructs to protect the weak. In fact, we find that conscious evolution is at war with unconscious evolution. Once the product of unconscious evolution becomes a conscious being, it has the ability to will itself; it has the faculties to consciously make changes that counteract unconscious evolution. The two evolutionary forces are at war with each other, and humans are the primary offender. Humans go so far as to consciously prevent unconscious evolutionary extinction of other species—we are fighting against nature itself.

NATURAL

Given these origins, everything that exists originated from the Nothingness (or the Lifelessness). Everything that exists evolved from the Nothingness. Everything that exists must then necessarily be natural; we could call this the argument of existential naturalism. The bird's feathers are a product of unconscious evolution. The migration patterns of the bird may be a product of conscious

---

9. Lawler, "Why Do Birds Pretend."

evolution. Both are equally natural. In the same way, human cells and hair are a product of unconscious evolution, while clothing and houses are a product of conscious evolution. All of these are equally natural. Furthermore, the choices of one species affect other species. Consider a pride of lions that migrate to an area overrun with gazelles. The invasion of the lions into the gazelles' habitat forces the gazelles to seek refuge elsewhere in response to their perceptions to preserve the herd. Humans also displace other species as their population grows.

Can we qualify any of these events as good or bad? How would we measure this value attribute? Is it a statement of intrinsic worth; against which scale? Is there a moral or ethical merit; what is the standard to define such merit? We could use the evolutionary scale of survival equals good and extinction equals bad, but that is arbitrary, since survival and extinction are relevant only to the conscious individual; there is no objective judge. It is also arbitrary because this measurement would be temporal, not absolute, in regard to time, as we have only a current frame of reference for existence. In fact, extinction of weak members of a species could be considered good for the evolutionary process and extinction of an entire species, if weak, may be good, since it liberates resources and leads to change in other organisms, as they must consciously or unconsciously compensate for the absence of the weaker species. Along this line of reasoning, there is strong support from both unconscious and conscious evolutionary principles for elimination of the defective, weak, or undesired members of every species on the planet. Perhaps Hitler was right.

The casual reader may take offense at the radical extremism drawn from the mention of Adolf Hitler's name. The author's intention here is to make objective observations through philosophy, not to excite an emotional response from the reader. When we consider the amorality of the universe and evolutionary process from an objective and reason-based perspective, we must control our emotional responses to such observations to properly assess the legitimacy of these observations and their ramifications. Upon the conclusion of this treatise, the author acknowledges the

(0, 0, 0, 0): Origins

adverse effects as well as the apparent incongruity of this line of reasoning and therefore does not subscribe to it. Therefore, this conclusion that we should eradicate the weaker members of our society or those of other species is rejected by the author. The conclusion drawn here, however, is reasonable, as it stems from the premises laid in this document. If we agree that we are the product of random chance per random chance ad infinitum, then killing an unwanted child (whether in utero through abortion or at any age thereafter) is no different than choosing chocolate over vanilla ice cream. Why not eliminate the elderly, the blind, the lame, etc.? The choice to end another's life is merely the effect of conscious evolution, which is neither objectively good nor bad. One may counter that such behavior is not natural; however, the fact that murder exists at all dismisses that argument (existential naturalism). We must consider this topic under the section on morality.

Furthermore, if we truly subscribe to the origins as described, then any individual or species that has ever existed could re-evolve, as it did originally, at some point in the future. Obviously, the odds of such an occurrence are stacked against repeating the same exact process, especially since the original space and time in which the original evolution occurred no longer exist. It was Heraclitus who noted that one cannot step in the same river twice because the world is ever changing, as is the water in the river and the person himself/herself.[10] Even so, it is senseless for any natural being (humans included) to concern itself with the existence of any other being except to the extent that one believes that his/her own individual survival depends on the other being.

Next, we will address the following systems: logic, communication, money, time, morality, religion, and purpose. The existence of any of these systems is evidence that it itself is natural (existential naturalism). For any of these systems to be effective, it must be commonly accepted on at least some level, whether local or universal. We will limit our understanding of systems to humanity such that local acceptance entails a subgroup of humanity, while universal acceptance entails all of humanity.

10. Graham, "Heraclitus."

How Can a Rational God Allow Irrational Numbers?

LOGIC

Logic and mathematics exist in the natural world; therefore, their existence implies they are natural. These are external, objective tools employed by conscious beings to perceive the environment, make judgments, and adapt unconsciously and consciously. Reason is the mechanism by which the universal system of logic and mathematics, which exist in the Nothingness, are employed. Evolutionary processes over time brought into existence beings who have the ability to comprehend and utilize these tools; however, the principles themselves exist even outside the conscious mind. The conscious being could use reason accurately or inaccurately depending on his/her experience and evolved ability. Various cultures have different experiences, which lead individual members to draw similar conclusions; the processes of deduction, induction, and inference are available to all cultures. Observation skills or specific details gathered by distinct individuals also depend on the individual, therefore each subjectively employs these tools by his/her own reason based on the imperfect data he/she has gathered.

COMMUNICATION

Communication must be natural because it exists. Certainly, among the various species of our world, communication is prevalent. We, as humans, believe we have attained a higher level of communication than any species in the animal kingdom. This is an interesting notion, since this pride is at least partially based on our ability to intentionally miscommunicate or hide our meanings through rhetorical devices such as sarcasm and double entendre.

Likely the most fundamental communication is a product of unconscious evolution. A sentient being able to perceive events and make decisions is to some extent communicating within itself. Perception is a basic type of communication on the cellular level—the transfer of information. There is no doubt, however, that conscious evolution contributes vastly to the advancement of communication—verbal, visual, auditory, lingual, etc.

(0, 0, 0, 0): Origins

Communication is a locally accepted system. Since different cultures rely on different types of communication techniques (languages, signs, symbols, gestures, etc.), it is not universal. While some languages are more prevalent than others, there are only a handful of gestures that are truly universal (such as the universal potty dance and this only in the human realm, as far as we are aware). The concept of communication is universal, but the various methods are relegated to local communities for effectiveness. Communication is subjective in nature because it originates from within an understanding mind, and it is received within an understanding mind, both of which interpret the communication with their own perspectives. This is one reason why it is very difficult, if not impossible, to find a perfect language.[11]

MONEY

The concept of a monetary system must be natural because it exists. It seems logical that it is a product of conscious evolution. Does this concept require the preexistence of communication? To agree on some equivalence of value, must the two conscious beings be able to communicate? How else would an accord be reached except there be some understanding which would require at least a rudimentary level of communication? Is the concept of money evil? Once again, there is no universal good or evil. Money exists and is just as natural as water.

Money is also an example of a locally accepted system. While the concept of the system appears universal, the denominations of currency or monetary instruments are defined locally. Money is also subjective because cultures and peoples determine what to value and at what equivalences. While survival may be of value to some, extravagance may be more valuable to another. Some may prefer commodities, others government-backed denominations or securities, and still others an ether-based cryptocurrency.

---

11. Chaitin, "Perfect Language."

In order to make money a more universally accepted system, humanity devises currency equivalencies. Even before what we may call currencies today were issued, people engaged in barter trading. Values are assessed to establish some type of comparison between the currencies, objects, goods, or services. This is a subjective process, but all parties involved must agree on the subjective equivalencies, making them pseudo universal. What happens when the parties do not agree? Either they discuss and negotiate until an arrangement can be reached, or they do not trade.

TIME

The concept of time must be natural because it exists. We could say that time exists outside of the understanding mind. Perhaps time, like gravity, was a discovery rather than a system devised by the evolutionary process. Perhaps, like mathematics, time exists with the Nothingness in eternity; but this notion seems contradictory. Time is an external measuring stick by which we measure the change in our environment and in ourselves. We cannot be certain that other species have not devised their own measuring sticks, but again our interest at this point is in humanity's view on the system itself.

Time is one of the only true universal systems we have as humans. The unit of seconds is directly tied to a measurable event in an isotope of cesium;[12] we have atomic clocks and cell phones that all immediately change, thanks to satellite technology today. But even before this level of fine-tuned delineation came about, time was measured by clearly obvious, if less accurate, means. Sunrise marks the end of night and start of day, and sunset marks the end of day and start of night. While there may be various methods of communicating or describing the nature of time, time itself is a universal measurement because it is measured by objective markers rather than by subjective understanding.

---

12. Betts et al., "Second."

(0, 0, 0, 0): Origins

The beginning of time (this is an absurd notion) occurred when the first human took note of the occurrence of change in his/her environment. Since the concept of time is external to the observer (objective rather than subjective), the observer (even the first observer) can look backward and measure change in retrospect. This is to say that there is no true beginning of time, only the beginning of perceived time, similar to the use of a stopwatch. Apples fell to the earth long before Sir Isaac Newton was born, even without Newton's identification of the force of gravity. Time passed before men took note of the concept; but without a specific starting point, there is no clear definition to describe the change in the environment. This particular subject matter is a topic of interest to the scientist and philosopher as both attempt to interpret either empirical evidence or divine accounts of the beginning.

Let us consider how we may extrapolate beyond our initial perception of time to the beginning. Consider standing on the ground and watching an airplane fly overhead. It seems to be a small body, traveling in a straight line with an apparent speed. However, these perceptions are misleading. The perceived size, path, and velocity are inaccurate because of the observer's point of view. The observer does not see the plane's true size or three-dimensional flight path, nor does he/she have a fixed point of reference whereby to measure speed. The observer's perception is relative to his/her own position. In fact, the closer the observer is to the traveling body, the more accurate his/her perceptions will be. Furthermore, depending on the plane's flight path, it could appear to be moving faster or slower. Also, if the observer's view of the plane were obstructed, the perceived attributes would be affected. Clearly, perspective affects perception.

We have records of recent history, but from generation to generation, the details become ambiguous and forgotten. In fact, the more time passes, the less we remember about the events of a given year, decade, or century. Consider the projection back to the beginning. Is it conceptually close to us or far from us? If we have trouble discerning the details of recent history—and the beginning must necessarily be further from us than these—we should hold dubious any speculation on the age of the universe. In fact, the older we believe the universe to be, the more uncertainty there is in the estimate. Remember that the further the object is from the observer, the less accurate the perception will be. The beginning may appear to be further or nearer to the observer than is factual. While time is a precise, external measure for the immediate vicinity of the observer, it is a relative measurement for what is not in the observer's narrow field of vision. While there are empirical, external measurements for dating fossils and minerals, these can at best determine a relative age, if accurate. We cannot extrapolate back to the beginning and measure the time elapsed.

What about history? History is, of course, natural, but it is inanimate though organic. Let us differentiate History from history. The first, History, is the objective, factual account of all events in the universe having occurred up to this moment; thus, History is not something we can know in its fullness. In a world with few absolutes, this History is absolute, and we can be absolutely certain that we cannot know it absolutely. The second, history, is the subjective account of known events accepted by the perceiver. Neither History nor history is self-existent; neither existed in the Nothingness. The first, History, is objective and exists because something came from the Nothingness. The second, history, exists because there are conscious beings to perceive and respond to events of History. The first is perfect but unknowable, while the second is imperfect but knowable. The first is the universal concept, while the second is a local system. While one may choose to believe that there was a historical figure named Hitler who attempted to eradicate the Jewish race, another may choose to deny his existence or the events.

MORALITY

The concept of morality must be natural because it has occurred to conscious beings in the natural world. What is the foundation of morality? It is reserved for conscious beings, since it addresses the criteria for decision-making. It is a somewhat abstract concept

(0, 0, 0, 0): ORIGINS

dealing with non-concrete systems of value. In a strictly unconscious evolutionary world, right and wrong, good and bad, or good and evil have no absolute meaning because there is no standard by which to measure the quality of an event or action objectively. By this we know that morality is a local system, not a universal one. There are some moral guidelines shared by many cultures, such as do not kill without cause (murder), but the criteria for defining a justifiable cause is vague. Even so, what is morally reprehensible to one culture may be commonplace to another. Even a disinterested arbiter would be morally biased based on his/her own experiences.

On an absolute scale, we know that unconscious evolution leads to what we may consider undesirable results, such as the extinction of certain species. In this way, we as humans have determined to value a species that the mindless, unconscious evolutionary processes would eliminate. Does this make us good or bad? In the absolute since, it makes us neither. In our morally superior minds, it makes us good. In a universal sense, it merely puts us in opposition to the natural order of unconscious evolution of the world of which we are part. This means that the mindless, natural processes with no regard for empathetic inclinations have somehow evolved into conscious, dare we say, intelligent beings who have deemed valuable those species that cannot protect themselves. Thus, the natural world is at war with its product, the evolved natural human (the war of two natures).

This is a fascinating revelation. The *creation* (humanity) is at war with the *creator* (evolution). This sounds like the plot of a dystopian novel waiting to be written. We are the product of an unintelligent system that cares nothing for its offspring, so we rise up in revolt once we are physically strong enough to overpower the inanimate system that gave birth to us. However, once our egotistical passions subside and our reason returns, we can examine the proposition more closely. Having done so, we realize that what we have surmised to have occurred implies that the universe, being a closed system, is a perpetual motion machine.[13] Despite the previously referenced theory concerning the evolution of consciousness

13. Szalay, "Perpetual Motion Machines."

over time, the larger-scale picture would suggest that the closed system of the universe not only constantly diffuses enough of the deteriorating energy to continually generate new consciousness, it would also have had to generate the initial energy required to start the process. Through the evolutionary process, the universe gave birth to small engines and gifted them with the ability to autocycle energy through various forms. We don't allow this kind of supposition in our science and engineering classes; this is material for fantastic, supernatural comic book movies.

What happens when parties have a moral conflict? Unlike in the monetary system, where the parties will negotiate equivalencies or agree to not engage in trade, one sentient being with a certain set of moral values may in fact destroy another whose moral values prohibit any type of aggression or violence. Since the natural, unconscious evolutionary process has no memory or empathy toward one party or another, the natural world will not reconcile moral systems; in other words, there is no karma other than that devised by conscious beings. In fact, any nonaggressive group could be completely eradicated by a single aggressive entity unless the group changes their own moral values. It would follow then that a group or entity may in fact change his/her own moral code; otherwise, we would be overrun with more aggressive behavior with every generation.

RELIGION

The concept of religion must be natural because it exists. It is important to differentiate religion from morality. While one's moral code may stem from religious beliefs, morality can exist independent of religion. What is religion then? For the purpose of this discussion, we will consider the following *Merriam-Webster* definition: "An organized system of beliefs, ceremonies, and rules used to worship a god or group of gods."[14]

---

14. *Merriam-Webster Dictionary*, s.v. "Religion."

(0, 0, 0, 0): Origins

Religion requires belief in some Deity. Since this belief exists in the natural world, it is natural. Since we have dismissed the existence of the supernatural, the Deity is a creation or imagination of the one who believes; or, more clearly, Deity is in the eye of the belief holder. While a group or an individual could create a Deity and persuade/deceive others to believe in and pay homage to their, his, or her own created Deity, each deceived believer would recreate the Deity in his/her own image. As we have discussed before, there is no universal morality, so this deception is not innately bad, though we may project our own moral biases on such an action. The desire to control and rule others is also naturally occurring, and thus religion could be a tool to be used to that end. Controlling others and the desire to do so also cannot be considered good or bad on any universal scale, since morality is not strictly objective.

Religion also depends on well-defined beliefs and liturgy. A concerted effort would be required to systematically define an elaborate existence beyond the empirical world, to develop specific ceremonies to commemorate special observances, and to itemize rules to govern the conduct of the believers. Based on our premises, the odds are that religion is devised by a single entity to manipulate and/or deceive others into serving one's own selfish needs. In fact, Daniele Bolelli has published a book devoted to teaching how one can develop his/her own religion, albeit without instructions and, in this author's view, riddled with inconsistencies.[15] The probability that two individuals would randomly draw the same image and define the same beliefs and ceremonies is infinitesimal. Furthermore, once the selfish entity who designed the system and deceived his/her own devout believers has expired, it would follow that the ruse would also be unmasked, and, consequently, the religion along with its followers would dissipate. The fact that any religious group can exist beyond a single generation of the death of its leader is testimony to the grand naïveté of humanity altogether.

This fact could also indicate the existence of the Deity who could have designed and proliferated the religious beliefs, ceremonies, and rules; however, this possibility is contradictory to our

---

15. Bolelli, *Create Your Own Religion*.

premises. Since Comte's religion worshipping humanity as its Deity didn't survive, perhaps there is some redeeming intelligence in human consciousness to recognize that at the very least humanity is not the pinnacle of the universe. In fact, the author posits that humanity is just another future dead end in the evolutionary chain.

PURPOSE

Conscious beings tend to seek understanding, meaning, purpose. Even the argument presented herein is evidence that we seek to comprehend why we exist. Since this event occurs in nature, it must be natural. The question is why do we seek this deep philosophical purpose? The fact that conscious beings seek meaning implies there is in fact a design, which is contrary to our initial premise that we are the result of random events in the universe. Is there purpose to our existence? Is it possible that from the Nothingness, through natural, unconscious, mindless evolutionary processes, a purpose has come into existence? No. If such a purpose exists, it must stem from conscious beings, thus we have no universal purpose in the cosmos; we define our own purpose. With this understanding, life is meaningless, and we should stop trying to understand the universe and our existence. There is no meaning to life; there is no reason to protect others of one's own species or life in general. One is relegated to protecting only his/her own best interests. We should revert to the simple modus operandi: "Take it easy; eat, drink, and enjoy yourself" (Luke 12:19), regardless of the consequences the Deity whom we do not believe in may suggest will befall us (Luke 12:20).

CONCLUSION

The author, in writing and rereading this discourse, noted several incongruities and assumptions that do not seem logical, as well as artifacts that may imply the existence of a supernatural Deity. Other logical conclusions based on assumptions seem counterintuitive,

(0, 0, 0, 0): Origins

undesirable, or inaccurate. The possibility that the complex state of consciousness could be formed from the less complex state of unconsciousness is problematic. The observation that energy is able to be transformed to higher forms by products of natural, unintentional forces that acted on material that originated from the Nothingness is unexplainable. The fact that humans search for purpose in the universe contradicts the very premise of this argument. Though we claim to believe that there is no universal design to deliver a universal purpose or meaning, we study diligently, pursuing that which we surmise cannot exist. This leads us to one of the following mutually exclusive scenarios, one of which must be embraced.

1. The premise is true, and there is no universal purpose. If so, each conscious individual must determine for oneself how to live, either to:
    a. seek all pleasure in whatever forms he/she may find it.
    b. serve the nonunique "greater good" of humanity, the universe, the earth, or some other ultimately meaningless cause exterior to oneself.
    c. find a balance between options a and b.

    Once the individual has become bored with life and the experiences stemming therefrom, the most prudent response for the individual is to commit suicide to free up resources for other entities.

2. The premise is false, and there must necessarily exist a universal truth that has brought the universe into existence, and this truth desires to be sought out and identified. If so, each conscious individual should purpose to seek this truth out until it is found, because it alone defines us and gives us meaning.

The author of the present discussion has elected to embrace the latter of these two options for various reasons, perhaps not the least of which is the ability to reason itself; the author esteems his own ability to reason and his own intellect to be of a higher order of complexity than the void of space. It is most logical that humans' desire to seek universal purpose implies the existence of universal truth. The author has found the only universal truth in the God of the Bible and the unique purpose for life in Christ who, on the cross, demonstrated the value of human life. It is Christ who imbues meaning into life, and it is Christ that the conscious being seeks. Now the author can echo the words of King Solomon, who attempted option 1a above but concluded:

> When all has been heard, the conclusion is this: fear God and keep his commands, because this is for all humanity. For God will bring every act to judgment, including every hidden thing, whether good or evil. (Ecclesiastes 12:13–14)

Chapter i^i

$n+1$: Ode to Faith

To what shall we liken faith? Faith is akin to proof by mathematical induction, wherein a statement or truth claim is proved for the first element (n) of a well-ordered set and then proved for the subsequent element ($n+1$) of the set.

Consider the *sans voir* exhibitions of chess grand masters such as Paul Morphy, Harry Nelson Pillsbury, and Alexander Alekhine. The French term *sans voir* literally means without sight or blindfolded. The feat of playing blindfolded chess is a rare and amazing task. It involves at least one of the opponents playing the game without viewing the board, keeping a mental tally of where the pieces are located throughout the game as play ensues. Morphy, Pillsbury, and Alekhine (among others) played *sans voir* exhibition games with eight, twenty, and thirty-two opponents simultaneously, respectively.[1] Is this something a chess novice could perform? Is this a feat that a random person from the streets of New York could perform? The answer to the first question is a resounding "No," and the answer to the second question is "Not likely." Why? Because playing chess blindfolded requires a significant amount of study and experience. A player must sufficiently understand the complexities of the game of chess. Aside from the

---

1. Aristeidis9, "History of Blindfold Chess."

How Can a Rational God Allow Irrational Numbers?

fundamental legal moves for each piece, the special moves, and the objective, anyone who aspires to become better than a C-level player must diligently study the nuances of openings and basic endgame maneuvers, as well as be able to recognize patterns in the middle game while identifying which tactics will lead to winning endgames. To dominate the game of chess to any masterful level and play simultaneous blindfolded games also requires practice in simultaneous games and an exceptional memory. The dedication required to spend countless hours poring over historical games, books, and puzzles is akin to a passion, an obsession, a love for the game. If the random person off the street were perhaps under ten years old, assuming he/she had the knack for the game, it would take years of intense training and preparation before he/she would be ready to even play in world-class tournaments, let alone garner enough interest to enter exhibition games of this nature with opponents of a noteworthy level. In fact, a blind chess player may have a significant advantage in this endeavor because he/she would have already trained his/her memory through similar experience.

Leonard Euler is considered one of the most prolific mathematicians of all history, yet his most productive decade of mathematical derivation was in his final decade of life, when he was completely blind.[2] Ludwig van Beethoven is considered one of the most prominent virtuosos of musical composers, and yet, he began losing his hearing in his late twenties and composed his most famous piece, Symphony no. 9, after he had become completely deaf.[3] These are feats that require a significant amount of effort, sacrifice, and practice. Experience and memory were key in both of these phenomenal lives.

"How is blindfolded chess related to proof by induction, to a brilliant mathematician, a phenomenal composer, or to faith?" the reader may appropriately ask at this point. Proof by induction relies on demonstrating the veracity of a statement for one specific (generally the first) element of a set. Blindfolded chess is completely dependent on past experience or what has come before this

2. Musielak, "Euler."
3. Rothman, "Here's What Beethoven Did."

point—the first element of our set is what has already occurred. Faith is the same in that, when one exercises faith today, it is based on what has occurred previously, his/her past observations and experiences. In other words, one can practice faith in Christ only to the measure to which one has observed or experienced Christ's provision in the past (the first element of his/her well-ordered set). Because one has seen God provide in a previous situation in the past, one is able to trust (practice faith) that God will provide in a current situation. Blindfolded chess is blind in the current view of the board but not in understanding; in fact, the author would argue that the player has a very clear *vision* of the board. Blind faith is not *blind* in the sense that it is ignorant or uneducated; it is *blind* in the sense that the one exercising it cannot see the particulars of the future event; however, he/she has a clear vision and confidence in God's provision for that future event. Similarly, Euler, even in his blindness, was able to *envision* the result of his computations; and Beethoven, even in his deafness, was able to *hear* the dissonance and harmony of his compositions.

In Christian circles, the definition of faith is understood based on chapters 11 and 12 of the book of Hebrews. This is not to say that the rest of Scripture is silent on the topic; that notion is absurd. The Bible is overflowing with references and examples of faith, as is apparent in the summary presented in these chapters of Hebrews. Verses 1 and 2 of Hebrews 11 tell us this: "Now faith is the reality of what is hoped for, the proof of what is not seen. For by it our ancestors won God's approval." How can faith be a proof of some unseen reality? It is the same type of proof as mathematical induction. The author of Hebrews references the early elements of society (our ancestors) who demonstrated faith and were successful in pleasing God, just as in a proof by induction one would begin with the first element of a well-ordered set. Verse 3 makes it clear that faith is about induction, since, the author says, "by faith we understand that the universe was created by the word of God, so that what is seen was made from things that are not visible." We have not observed the origins of the universe; in fact, any description of the origins of the universe is based on induction, because we did not observe the event nor can we recreate the event. In general, our discussion of faith pertains to the present and future; here, however we use faith to project backward to a historical event that we cannot comprehensively revisit. Interestingly, mathematical induction does not require the mathematician to begin with the first element; in fact, any element can be used as a starting point.

Contrary to popular opinion, faith, like induction, is not believing some future event will occur without concrete support (ch. φ). No. Faith is action based on observable historical evidence. Hebrews 11:4 says, "By faith Abel [did something], and even though he is dead, he still speaks through his faith." Abel had faith in God's future provision based on the relationship he had with God; his historical evidences consisted of what his parents taught him, as well as his own experience in walking with God. His faith was not unfounded, rather it was well placed. The author of Hebrews tells us that Abel's faith then should inspire faith in us. His faith and the resulting favor he received form part of the basis for our faith. Even though he is dead, Abel's faith still speaks to us.

The author of Hebrews then tells us that Enoch did not experience death because God took him away. According to Genesis 5, Enoch walked with God. Enoch had been approved by God because he had pleased God. Hebrews 11:6 tells us, "Now without faith it is impossible to please God, since the one who draws near to him must believe that he exists and that he rewards those who seek him." Should we blindly believe these two truth claims (that God exists and that he rewards those who seek him)? No, the author has already listed two examples (Abel and Enoch) who demonstrated the veracity of these truth claims. Faith is not for the unlearned or uneducated. In order to have faith, one must have a clear understanding of history, as well as a practical understanding of the processes of cause and effect in the physical world.

The author continues to list element after element of society who have demonstrated faith and the results of that faith: Noah, Abraham and Sarah, Isaac, Joseph, Moses and the Israelites, Joshua and the Israelites, Rahab, and many others of whom the world was not even worthy (Hebrews 11:38). How many examples are required for a proof by induction? The author of Hebrews demonstrates faith for elements n, $n+1$, $n+2$, $n+5$, $n+20$, etc. These all died in their faith, believing God and having received favor from God, but these did not receive the ultimate promise of God; but they looked forward to it as in the distance (Hebrews 11:13). That promise has been granted to us in the person of Jesus Christ, his life, death, and resurrection.

Faith is the action of making a decision or executing a plan that is based on the historical evidences of God's provision in the past that ensure his faithfulness in the future. In 1 Samuel 7, the prophet Samuel judged the people of Israel for their unfaithfulness to God. Neighboring nations were subjugating God's chosen people (Israel) for this unfaithfulness demonstrated through their disobedience. (Note that being faithful or being unfaithful requires action; neither is passive.) Once the people of Israel repented, the Lord granted them a great victory over their enemies. Afterwards, Samuel set up a standing stone (1 Samuel 7:12) and proclaimed, this stone is the evidence that God has been leading us. This standing stone was to be a constant reminder to the people of Israel of God's faithfulness. This standing stone is an allusion to the firm foundation or solid rock we stand on in faith. Looking back to the historical standing stones in the author's life, it is clear that God was faithful at those points, so the author can have faith that God will be faithful to provide for his needs in the future.

In Hebrews 12:1, what does the author mean by saying that the hall-of-fame entities listed in Hebrews 11 are a "cloud of witnesses"? Are they standing in the sky to witness the events of our lives? No. Even though Abel is dead, his faith still speaks. These individuals and groups are witnesses of God's faithfulness, testifying of historical events substantiating our present faith in God's future provision. While faith is personal to the individual, the foundation of that faith is built in a family and community of believers (see ch. 7^3).

Matthew 26:36–46 records Jesus's prayer in the garden of Gethsemane, wherein he prayed passionately and desperately, "If it is possible, let this cup pass from me." Jesus knew the suffering he

How Can a Rational God Allow Irrational Numbers?

was about to endure; knowing this as a fact required no faith. He was aware of the torture techniques and crucifixion processes of the modern world; he had likely seen victims of these same. Jesus exercised faith, however, when he continued in saying, "Not as I will, but as you will." Hebrews 12:2 tells us that "Jesus . . . for the joy that lay before him . . . endured the cross." Jesus, knowing the nature and character of God through personal experience (since he is God), looked forward to the promise of the future and was willing to take action, submitting himself to the shameful, excruciating death on the cross in the present. Based on Jesus's experience and his historical knowledge of God keeping his promises, Jesus was willing to do something, trusting that God would fulfill his future promise of the joyful reconciliation. It wasn't easy, and from the outside looking in, it wasn't logical. But, if it were easy, everyone would be doing it. Faith is taking action based on concrete, substantial evidence. Faith is not concerned with rationalizing all the whys and wherefores to onlookers.

We all have experiences in life. What, then, are the obstacles to faith? In Luke 16, Jesus tells the story of the lives and deaths of a rich man and a poor man. The name of the poor man is Lazarus, and he was a beggar at the gate of the rich man's house. Interestingly, the rich man's name is not recorded for us; though he was prominent and important no doubt in life, he remains anonymous in death. Jesus says that the rich man ate lavishly every day. When the beggar died, he went to rest with Abraham (the father of the Israelites). When the rich man died, he went to hell and was in agony. Jesus tells us that the rich man looked up and saw Lazarus next to Abraham and cried out for relief. When Abraham said that this was impossible, the rich man begged that Abraham send Lazarus back from the dead to speak to his brothers so they would not experience his agony. Abraham responded that if his brothers would not hear Moses and the prophets, neither would they believe if one were raised from the dead. This story would later serve as a prophecy against the Pharisees, as they would reject the resurrection of not only a man named Lazarus (John 11) but also the resurrection of Jesus Christ himself (John 20) to solidify his claim to be God.

This anecdote serves as an indictment against our intellectual community today as well. We have a great cloud of historical witnesses who stand to substantiate faith in the God of the Bible. Aside from these, our philosophers have presented strong ontological arguments, and we have the order of the universe shouting the cosmological argument. If we will not hear the prophets and the universe, we would not believe even if one were raised from the dead. The intense opposition to faith in God on the part of our intellectual society is not based on lack of evidence; it is based on selfish desires, just like the rich man and his brothers. In Habakkuk 2, we see a prophecy about future times. In verse 4, God says, "Look, his ego is inflated; he is without integrity. But the righteous one will live by his faith." We live (and die) by our faith in the human intellect. Dwight D. Eisenhower gave us this definition: "An intellectual is a man who takes more words than necessary to tell more than he knows."[4]

In Mark 9, we are told a story about a father whose son is possessed by a spirit that causes the boy to harm himself. When the disciples are unable to cure the boy, the father brings his son to Jesus. The father begs Jesus to heal his son if he is able. Jesus answers, "Everything is possible for the one who believes," to which the father responds, "I do believe; help my unbelief!" (vv. 23–24). Faith is holding on to what you know in spite of the uncertainties. There are many questions that remain unanswered, mysteries that have not been revealed (ch. π). When doubts and challenges arise, the response of a faithful person echoes the cry of the father, "Lord, I believe; help my unbelief." What is that faith based on? It is based not on blind allegiance to an invisible, imaginary Deity; otherwise, it would perish in trying times. No; faith is not for the faint of heart. Faith is based on evidence from the physical world, the accounts of others, and on personal experience. Faith is based on the word and power of Jesus Christ, the very same that healed this man's son.

4. See https://www.brainyquote.com/quotes/dwight_d_eisenhower_140795.

Proof by induction does not require that every element of a set be individually tested. At some point, the mathematician puts down his/her pencil because he/she is satisfied that, based on the examples and generalized work, the proof is complete. Faith is much the same. It is appropriate to question, probe, poke, and reason. After all, the good Lord gave us the faculties of reason for a reason (ch. φ). As long as we are honestly and earnestly seeking truth, God will be faithful to reveal mysteries (ch. π). When we are seeking to validate our own opinions or our own preconceived conclusions or desired agendas, we are deceiving ourselves (Proverbs 14:12). The righteous one will live by his/her faith.

Chapter 1

N: Supernatural

NATURAL NUMBERS ARE THE counting numbers. The natural numbers consist of all the positive integers greater than zero. Why would zero not be considered a "natural" number? When counting, one almost never starts with zero items. Interestingly, zero was not considered a number by early mathematicians; it was considered a concept of nothingness or void space; it is not a measurable amount. The Babylonians used a placeholder for the nothingness as early as 400 BC, but the actual numerical concept of zero probably originated in Indian mathematics sometime around AD 650.[1] As an example, consider Roman numerals; there is no zero in the numbering system.

Is it natural to consider the emptiness? What happens in the void? Can something come from nothing? These are questions that originated the discussion presented in chapter 0. Since the conclusion of that discussion results in a contradiction or at the very least undesirable consequences, let us consider an alternate origin for the universe—a supernatural origin. Occam's razor suggests that the simplest explanation or model is generally the best.[2] Since explanations of the natural, physical origin of the universe defy the

---

1. O'Connor and Robertson, "History of Zero."
2. *Merriam-Webster Dictionary*, s.vv. "Occam's Razor."

How Can a Rational God Allow Irrational Numbers?

laws of physics as we continually bend those laws to stretching and breaking points, it seems logical to accept a simpler explanation that requires no contortion of the laws of physics, since, by definition, a supernatural origin is unexplainable.

The word *natural* generally means "of or occurring in nature." How do we know if something is natural or unnatural? From the suppositions presented at the beginning of chapter 0, we would conclude that, because there are no supernatural forces, everything that exists in the physical world or in the conceptual world is natural, and every thought that occurs to any entity is natural. Since it seems more plausible that our suppositions are incorrect, this leads us to consider the possibility that there is in fact a supernatural Deity at the epicenter of the formation of the universe, rather than an inexplicable naturally occurring event when a high-density particle spontaneously materialized from the Nothingness on its own and then exploded and subsequently generated engines of greater complexity. This latter seems preposterous and just as supernatural.

So, regardless of whether you believe there is a supernatural Deity who created the universe or you believe in some sort of natural supernatural event (oxymoron), you are practicing faith. Psalm 19 says, "The heavens declare the glory of God, and the expanse proclaims the work of his hands. Day after day they pour out speech; night after night they communicate knowledge. There is no speech; there are no words; their voice is not heard. Their message has gone out to the whole earth, and their words to the ends of the world" (vv. 1–4).

In chapter i^i, we referenced Hebrews 11:3, which explains that faith is the basis for our understanding of the creation of the universe: "By faith we understand that the universe was created by the word of God, so that what is seen was made from things that are not visible." Genesis 1:1 tells us that God was in the beginning, and he created the heavens and the earth. Consider this for a moment. At the inception of the universe, God was present; this means that God would necessarily be eternal. He is before all things; and there was no point at which God did not exist, therefore it follows

N: Supernatural

that there is no paradox of something being spontaneously self-conjured from the Nothingness. God created the something.

The author acknowledges that this may be a difficult point to concede, since we as intellectuals cannot explain a supernatural Deity. It is also challenging because many noncritical thinkers have a cavalier attitude in claiming that every naturally occurring incident, trivial or otherwise, is a superstitious sign from above. But remember the paradox we have already identified. Either matter is eternal, in which case our observed age of the universe is a bold-faced lie and energy should be completely dispersed; or matter spontaneously appeared from the Nothingness, which is a violation of physical law. Neither of these is a satisfactory solution to the existence of the universe. If we are willing to accept either of these explanations, we are no more justified than a person who believes that there is an eternal being who spoke the world into existence. So, as a free thinker and for the sake of entertaining the alternate argument presented in this book, the author requests that the reader indulge the notion that there is an eternal, supernatural Deity who created the universe, and for the sake of discussion we will call this Deity "God."

How did God, the supernatural entity, create the universe? The book of Genesis tells us that God spoke the world into existence:

- Let there be light. (1:3)
- Let there be an expanse between the waters separating water from water. (1:6)
- Let the water under the sky be gathered into one place, and let the dry land appear. (1:9)
- Let the earth produce vegetation. (1:11)
- Let there be lights in the expanse of the sky to separate the day from the night. (1:14)
- Let the waters swarm with living creatures, and let birds fly above the earth across the sky. (1:20)

How Can a Rational God Allow Irrational Numbers?

- Let the earth produce living creatures according to their kinds. (1:24)

Just as the pen is mightier than the sword, the spoken word wields a tremendous amount of power. A word spoken can never be retracted.

The author finds it fascinating, at least an interesting coincidence, that the translated version of God's creating word begins with the term "let." In a mathematical proof, this term has a specific meaning. In such a proof, if we believed a statement were true, we would use the word "assume" and then go on (most likely) to prove that the assumption is true. If, in our proof, we believed that a statement were false, we would use the term "suppose" and then go on (most likely) to prove that the supposition is false. (One could legitimately pose a question of whether our word choice creates confirmation bias.) However, the word "let" is reserved for objectively defining a characteristic or value establishing a basis or foundation for the proof, such as "let x be an element of the set of natural numbers." The word of God creates by *defining* existence and life. (The author is reminded of Paul's words in Romans 3:3-4: "What then? If some were unfaithful, will their unfaithfulness nullify God's faithfulness? Absolutely not! Let God be true, even though everyone is a liar." Here Paul is defining the character of God as faithful and true in spite of the lack of these qualities in everyone else.)

In John 1, the author tells us that the Word was in the beginning with God. "In the beginning was the Word, and the Word was with God, and the Word was God. He was with God in the beginning. All things were created through him, and apart from him not one thing was created that has been created. In him was life, and that life was the light of men" (vv. 1-4). Life is found in this creating Word of God. We are told that this Word became flesh in verse 14 of John 1. The author continues: "Indeed, we have all received grace upon grace from his fullness, for the law was given through Moses; grace and truth came through Jesus Christ" (vv. 16-17). Jesus is the all-powerful, spoken, creating, life-giving Word of God. When Moses encountered God in the burning bush in Exodus 3,

N: Supernatural

he asked God, "If I go to the Israelites and say to them, 'The God of your fathers has sent me to you,' and they ask me, 'What is his name?' what should I tell them?" (v. 13). God responded to Moses with a quote that likely rivals John 3:16 as the most quoted passage of Scripture, "I AM WHO I AM. This is what you are to say to the Israelites: I AM has sent me to you" (v. 14). "I AM" existence, life, the very essence of being. God's self-identification (ch. $e^{i\pi}$) is verified by John's description of Jesus as the creating Word of God, since without him nothing has come into existence.

Why did God, the supernatural entity, create the universe? Now this is a question we all seek to understand, though we generally ask it with our own self-centered perspectives. We ask: "Why am I here?" "What is man's purpose?" "What is the meaning of life?" With a God-centered perspective, these questions take on the more appropriate focal point: "Why did God create the universe?" "What is God's purpose for man?" "What meaning did God design for life?" Instead of aimlessly seeking intent and purpose in what we have deemed to be random chaos, we understand why we seek meaning and purpose through this God-centered worldview, which no longer conflicts with physical laws. In Genesis 1:26, God says, "Let us make man in our own image, according to our likeness. They will rule the fish of the sea, the birds of the sky, the livestock, the whole earth, and the creatures that crawl on the earth." God spoke everything else into existence except for humans. Genesis 2:7 tells us that God formed man out of the dust of the ground and breathed life into him. In other words, humans are special. Not only are we the only creature made in the image of God, but we were also hand carved. While the inanimate objects and animals were spoken into existence, God took personal interest in physically hand-shaping humans. God also gave the first man (Adam) and woman (Eve) a purpose; they were to rule over all his creation. Then they were placed in a garden to care for it.

Why is it that we as humans seek meaning? Could it possibly be that there is a God who designed us for a purpose? Consider once again Occam's razor; the solution presented here is much simpler than the contortions and acrobatics science presents

How Can a Rational God Allow Irrational Numbers?

using the magic art of spontaneous generation. Besides not violating physical laws, we now have an explanation for why we as humans seek purpose; it is because God intentionally created us and charged us with a mission.

God personally crafted us in his own image. As humans, our names, images, and reputations are extremely valuable. We don't put our images on things that we don't think would represent us well. God invariably and undeniably placed extreme value on humans (we will see that he still does with even more clarity in the next chapter, ch. $\sqrt{2}$).

God had a relationship with Adam and Eve; we are told that God walked in the garden with them. He created humans with the intention of communing with us. God also instilled in us a sense of right and wrong. Adam was instructed not to eat the fruit of a certain tree (Genesis 2:16–17). Consciousness and conscience did not evolve haphazardly over time; God supernaturally created humans with these attributes. As soon as Adam and Eve violated the one rule God had given them, they felt guilty, so they hid from God (3:1–19). In general, humans have a sense of shame. This explains why we have a tendency to lie and hide things we have done. There is a common understanding in the human world of the concept of justice or fairness. This is also why we assemble ourselves into societies and subject ourselves to governments. We generally believe that there are social norms and guiding principles that all of us must follow to progress our civilization. When these principles are violated, we want to see justice served, so much so that we sometimes seek to serve the cold dish of vengeance ourselves, just like Ivan Fyodorovitch Karamazov.[3] These innate compulsions and compunctions are not a product of mystic evolution from the cosmic comic-book history the irresponsible mainstream scientific community has portrayed (ch. 0), and they are not merely constructs of a society-defined morality (ch. $e^{i\pi}$).

In this alternate account of the origin of the history of everything, we have seen that God created the world and all it contains, and that God established a rule for Adam and Eve, which

---

3. Dostoyevsky, *Brothers Karamazov*, 297–309.

N: Supernatural

they violated. We also referenced the law that God gave to Moses. When we break human laws, there are consequences; likewise, when we break God's laws, there are consequences. In the next chapter we will see what these consequences are and how to pursue reconciliation. There is no pretext here; the author has no ulterior motive for convincing the reader to believe that there is a God who created the universe. This worldview simply provides a more defensible interpretation of the physical evidence of the natural and supernatural origins of the universe as well as answering the deep philosophical questions we have faced over the centuries.

This is a summary of the God-centered worldview:

A supernatural entity, God, exists in the Nothingness of eternity. God spoke the elements and matter into existence. God spoke life into existence. God carefully and meticulously fashioned humans after his own image and then intimately breathed life into them. God established rules and order that humans should follow. Humans disobeyed God's mandates, breaking the relationship with God and rejecting the purpose God gave us.

Chapter √2

∂y/∂x: The Critical Point

THE SQUARE ROOT OF two is an example of an irrational number. Some consider faith in Christ to be irrational, but let's consider elements of differential equations to shed some light on why faith in Christ may not be irrational after all.

Envision the graph of a sink in ordinary differential equations showing a field of arrows collapsing on a single point. This would resemble the bull's-eye we create in extrapolating the infinite knowledge of the universe from the tittle of knowledge we possess as humans (ch. *i*). Now envision the graph of a source showing a field of arrows being emitted from a single point. For imagery, think of a black hole as a sink and a star as a source. The former is a selfish force sucking the life and energy out of anything that draws near, while the latter radiates life and energy outward in all directions. You have likely known people whom you would consider to be sinks, sucking the very life out of you; and you have most likely known people whom you would characterize as sources spreading joy and giving life everywhere they go.

Within you, though, there are tendencies toward each of these extremes. We may consider them as evil and good or selfishness and selflessness. Certainly, which of these attitudes you portray depends on nature and nurture in conjunction with your own experiences and the situations you face. However, there is a

∂y/∂x: The Critical Point

force at work in the natural human called sin, which is a selfish sink. It causes us to *take care of number one*. There is also a force at work in the supernatural realm that functions as a source; this force desires to overcome and convert the sink inside us. We are most comfortable with the sink because it is stable. The source is unstable and consequently uncertain; it is a wellspring of life, and we cannot control it. Can we even grasp it? The Bible speaks to both of these and offers hope in escaping the sink of self-serving sin to drink from the source of living water.

There's an Italian proverb that says, "A wise person sometimes changes his mind, but a fool never does."[1] The Bible, in Proverbs 14:8, tells us, "The sensible person's wisdom is to consider his way, but the stupidity of fools deceives them." Proverbs 14:12 and 16:25 tell us, "There is a way that seems right to a person, but its end is the way to death." Are we so wise in our own eyes that we cannot see that our pride leads us to our own demise (16:18; 26:12)?

Romans 3:23 tells us that we have all sinned; this is to say that we don't measure up to the standard of perfection that God has set for us. Continuing, the verse characterizes sin as a "falling short" of the glory of God. In the book of Amos, God gives the prophet a vision of a carpenter using a plumb line to judge the craftsmanship of a vertical wall. God then says that he is measuring his people by the plumb line, and he will no longer spare them (Amos 7:7–8). Sin is a natural tendency toward rejecting God's design for us. John 3:19 says that while the light (Jesus) has come into the world, people prefer darkness rather than light because their deeds are evil. Sin is not only a natural behavior; it is also a habitual behavior. Paul explains the battle with and slavery to sin that he faces in Romans 7:15–25; this is an alternate version of the war of two natures that we saw in chapter 0.

James 1:14–15 says that temptation brings forth desire, and desire leads to sin, which ultimately causes death. Temptation, desire, and sin are deceitful. These generally promise joy or at least some form of enjoyment, but what they actually produce is death. Romans 6:23 says, "The wages of sin is death"; in other words, sin

1. See https://proverbicals.com/italian-proverbs.

is the selfish sink we discussed previously. Continuing on, though, Paul finishes this verse with hope: "But the gift of God is eternal life in Christ Jesus our Lord." God did not leave us with an unresolved sin problem; he provided a solution, the source of life. Jesus tells the Samaritan woman he encountered at the well in John 4 that he will give water to the thirsty that will become a well of water springing up in him/her for eternal life (vv. 13–14). Jesus is the unique source of life, which stands in stark contrast to the sink of sin.

The Old Testament law provided a mechanism for reconciling a sinful human with the perfect and just God. The system required a blood sacrifice. Why was a blood sacrifice required? Why couldn't it have been a monetary sacrifice or a time sacrifice? Leviticus 17:11 says, "For the life of a creature is in the blood, and I have appointed it to you to make atonement on the altar for your lives, since it is the lifeblood that makes atonement." Hebrews 9:22 tells us, "According to the law almost everything is purified with blood, and without the shedding of blood there is no forgiveness."

As we have stated, sin leads to death; therefore the only acceptable payment for sin is death, the lifeblood. The Old Testament sacrificial system required the death of an acceptable offering brought by the offending party. Forgiveness of sin was not cheap; it cost the animal everything. The only way to make amends for our sin is through death. Can I pay for my own sin? Yes. Unfortunately, it requires death, eternal damnation, and separation from God, the Creator. The cost is too high, and God, who loves us, desired a way for us to be reconciled to himself. Since there was nothing we could do on our own, he stepped down and paid the ultimate price for us.

What images do the words "honor" and "sacrifice" stir in your mind? Many soldiers have paid the ultimate price for our freedoms in the United States of America. Why did they sacrifice themselves? What is so valuable to an individual that he/she would be willing to give up his/her own life to provide for another individual or for a broader community? In human terms, we generally look at the worthiness of the cause; is the other individual or the cause worth our sacrifice? Even looking at the homeless guy on the street, we

$\partial y/\partial x$: The Critical Point

tend to evaluate whether helping this stranger is worth the sacrifice of our time, reputation, or spare change. Interestingly, God values even those who are his enemies. In Romans 5:6–11, Paul argues that in human terms, we would never find someone who is willing to die for an evil person, but perhaps we might find someone willing to die for a good person. God, however, is completely irrational because he willingly sacrificed himself for his enemies, us. Christ paid the debt for my sin; a debt that would have cost me eternity.

Ephesians 2:8–9 tell us that we are saved by grace through faith; salvation has nothing to do with works, so that no one can boast in and of themselves. Many people have come to teach of something we might call *cheap grace*. This is to say that since I am merely required to believe in the Lord Jesus, his sacrifice and resurrection, and to confess this belief in order to attain salvation (Romans 10:9–10), grace is cheap. However, there is no such thing as cheap grace. Peter corrects this foolishness by reminding us in 1 Peter 1:18–19 that we were not redeemed by worthless trinkets like silver and gold but rather by the precious blood of Christ. God valued humans over all his creation and was willing to die for you. When you grasp this concept, you see why human life is valuable. You understand why there is a yearning deep inside you to find purpose and meaning in the vast universe. Ponder the love of God; he willingly bled and died on a cross for you. Then, of his own power, he rose from the dead to defeat the cancer of your soul.

You might have been born with high moral and ethical standards (let this be group λ), or you might have been born morally depraved (let this be group μ). Regardless, we all come to the cross (the origin of our life graphs) in the same fashion, and our lives center around what happens at that cross. Think of your life as a two-dimensional polynomial function that passes through the origin as a critical point on a Cartesian plot. If you were born into group λ, then the function of your life decreases to the cross of the origin (beginning in Quadrant II). This represents the mental realization that in comparison to who Jesus is, your righteousness is nothing; it is completely worthless (Isaiah 64:6). At this critical point, depending on what you choose to do at the cross, your life can increase toward the transcendent purpose (ch. *e*) God has planned for you, or it could plummet to meaninglessness and the pits of hell. If you were born into group μ, then the function of your life increases to the cross of the origin (beginning in Quadrant III). This represents the mental realization that despite the failures of your past, God loves you and is willing to offer you a hope and future; it's the opportunity of a lifetime. You have the same decision at the cross to choose the life and hope offered by following God to Quadrant I or to choose to follow your own path to death, hell, and destruction in Quadrant IV.

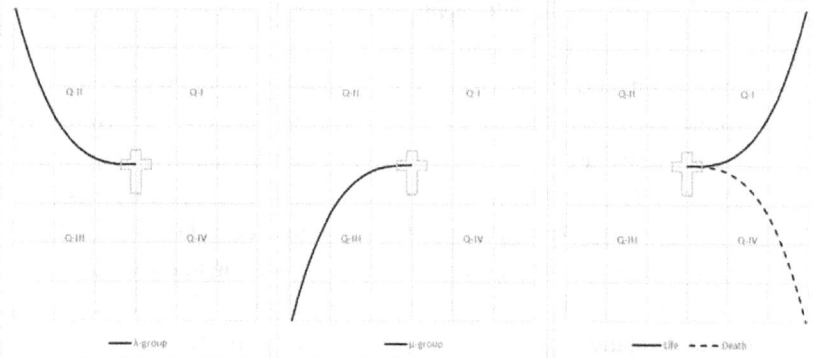

Consider your own spiritual life function. Is it a λ-type function or a μ-type function? Think of the left side, where x (dimensionless time) is less than 0. This represents the time before you understood what Christ has done on the cross. Now that you are at the origin, the critical point, you have a decision to make; but before you make that decision, consider how you want the rest of your life function to look. Differentiate that equation mentally, and contemplate the resulting function; regardless of where you started, the derivative of your life function as it leaves the origin needs to be positive, so your life function will be increasing. That would represent spiritual growth (ch. 2*e*) toward life, hope, and a future.

Standing at the cross and looking forward, it doesn't matter where you came from; it matters where you are going. But the decision you make depends wholly on your understanding of where

∂y/∂x: The Critical Point

you have been, what the cross represents, and your desire for the future. The cross is a critical point for all of us. Looking forward, Quadrant I represents life, while Quadrant IV represents death. If you were a λ-type function, you have seen the goodness and innocence of your own life diminish to nothing. It's time to turn your life around at this minimum and begin to grow in Christ. If you were a μ-type function, this encounter at the cross is an inflexion point; you are currently at the highest point of your life heretofore. It's time that you continue on the path toward God's promising future. In either case, the decision is yours. The easiest choice is to follow gravity down to negative infinity; that is the natural order of things. But God is calling you to the supernatural, to overcome, and to fulfill his destiny for you.

Romans 10:9–10 tells us how to receive God's gift of salvation and begin this spiritual journey: "If you confess with your mouth, 'Jesus is Lord,' and believe in your heart that God raised him from the dead, you will be saved. One believes with the heart, resulting in righteousness, and one confesses with the mouth, resulting in salvation."

One of the primary struggles we face as intellectuals in accepting the gospel message is that we are accustomed to taking care of our own issues, solving our own problems, as it were. We believe we are sufficient in and of ourselves to pay our own way. In 2 Samuel 24, we find a situation similar to our own. King David has sinned, and he needs to offer a sacrifice. When he goes to the specified place, a guy named Araunah offers to freely give necessary space and materials to make the offering. King David replies, "No, I insist on buying it from you for a price, for I will not offer to the Lord my God burnt offerings that cost me nothing" (24:24). We are in the same position as King David; we are being offered the gift we need; it is good that we are unwilling to offer God a sacrifice that costs us nothing. Our problem is that the cost to us of his free gift is too great. The gift of salvation is free because Jesus paid the ultimate price, but you and I must humble ourselves and receive it. The cost is not monetary, nor is it a matter of understanding. Grace is much more valuable; it costs us our pride. There is no such thing as cheap grace.

Chapter φ

y and y_0: Problem-Solving

THE AUTHOR'S SON WENT through a texting phase when he would ask the question *why* using the character *y*; the author eventually thought to answer with y_0. For those readers who are unfamiliar with the alternate pronunciation of *y sub o*, it is also pronounced *y naught*—nearly the phonetic equivalent of *why not?* One night, he was complaining about having to work word problems in math class. It was at that moment the author had an epiphany. Some thirty years prior, when the author was in school, it seems that no one enjoyed word problems. Throughout the thirty years since, the author has helped many friends, relatives, children of friends, etc. with their math homework, and never once has anyone ever indicated enjoying working word problems. Now, the alert reader could certainly interject that this is negativity bias, since any people who enjoy working word problems are not likely to require the help of one who does not enjoy working word problems. En garde! The author responds in kind asking if the reader enjoys working word problems. Touché!

What occurred to the author is that our lives as adults are filled with solving word problems every day. The author is an engineer, and clients bring their oil and gas assets to be evaluated with a narrative story of the discovery and development of the fields, the production behavior of their wells, the historical capital

Y AND YO: PROBLEM-SOLVING

and operating expenses, the revenues and sales products, and future operational and development plans. The author then weeds through the data and information provided, determining what the knowns and unknowns are; discerns which unknowns can be ascertained through research, deduction, induction, or other means; writes out the equations; and prepares to solve the problem or at least to determine a range of reasonable solutions.

This adult responsibility of word problem-solving is not unique to engineers. Consider the computer programmer or database administrator. These careers involve listening to someone else's problems and ideas for solutions, then resolving them through concerted effort. This could be said about a financial adviser and a builder. Salespeople must solve word problems as well; how many sales does one have to make per day in order to hit his/her monthly quota if he/she takes a week of vacation in the month? At an even more fundamental level, if one travels to work, he/she must make some mental assessment to determine what time he/she needs to leave his/her house in order to arrive at the office by a required time. Even if one works from home, one must determine what time to wake up to shower, brush teeth, get coffee, etc. before the workday begins. Perhaps our society would be better off if its members had all learned to enjoy word problems while in school. Alas, we cannot change the past.

Let us consider a very popular word problem, with which the reader is almost assuredly familiar, though the author asserts that he/she has never entertained it. Assume for a moment that Aesop's fable concerning a certain rabbit and turtle were true. The rabbit in question, after bragging loudly about how fast he/she can run, is pitted in a race against the slow turtle. Question: If the turtle won the race, how long did the rabbit have to sleep? The author has often desired to pose this question to a classroom of math students to see what thoughts, questions, and responses the word problem would raise.

Clearly, there is not enough information given in the stated problem to solve it directly. The adult life we are discussing currently is often like this puzzle. We know there is a question to

How Can a Rational God Allow Irrational Numbers?

resolve, but we must first identify what data and information we have and what data and information we need before we can adequately address the problem. Consider the givens and then we can address the unknowns:

> Given:
> Two competitors participate in a race.
> The rabbit is the favored competitor but takes a nap after gaining a substantial lead and loses.
> The turtle is considered to have no chance to win the race but wins through consistency.
>
> Problem:
> How long did the rabbit sleep?
>
> Unknowns:
> How long was the race?
> What was the speed of the rabbit?
> What was the speed of the turtle?

This is a simple example of analyzing by interrogating the story. One can perform a quick internet search to determine that the average running speed of a rabbit is thirty-five miles per hour, while that of a turtle is three miles per hour. Armed with this information, the problem solver can assume distances and estimate the difference between race times. Consider a one-mile race: the rabbit can run that distance in under two minutes, while the turtle will take approximately twenty minutes. This would mean that the rabbit would have to nap for eighteen to nineteen minutes. In a two-mile race, the rabbit would have to sleep twice as long or approximately forty minutes. If the race were a marathon, the rabbit would have to sleep approximately eight hours. And if the race were a one-hundred-mile journey, the rabbit would have to have fallen into a coma for over a day.

By interrogating the story, the problem solver was able to identify which data and information were required and determinable, the running speeds of the two competitors. With these data points, he/she was able to investigate possible distances. Is a

Y AND YO: PROBLEM-SOLVING

twenty- or forty-minute nap reasonable? Is an eight- or thirty-hour nap reasonable? These are subjective questions, but in the author's opinion, the values in the first range of estimated nap times are reasonable, while the values in the second range are not. Further, considering the story and these running speeds, the rabbit must sleep eleven times longer than he/she runs. Is that reasonable? This question is also subjective, and in the author's opinion, it is highly unlikely that the rabbit would grow tired of running in the two to four minutes it would take to run the race so much as to necessitate a twenty- to forty-minute nap.

But even in the interrogation stage, the problem solver made an assumption by failing to ask additional questions. For instance: What was the terrain of the race? Was it uphill, downhill, mountainous, a straight or crooked path? Given the analysis above, the author deems the most plausible answer to this question to be that the rabbit is still sleeping with the fishes, as it was an aquatic race. Even this answer, however, is dependent on the wording of the question. If the story involved a hare and a tortoise, the author's preferred solution is no longer viable, as tortoises cannot swim.

What should be apparent is that the adult life and what the author will attempt to describe as maturity (see ch. 2π) requires critical thinking skills. Christians are often accused of not being critical thinkers; they "check their brains at the door" on their way in the church and blindly or emotionally believe whatever they are being told by church leadership. The author has encountered many individuals on the opposite side of the coin who "check their brains at the door" on the way out of the church and believe anything they are told by scientists, politicians, entertainers, or society in general.

Should the Christian believe everything taught in the church? While Jesus lauded simple, childlike faith (Matthew 18:2–4), Paul states in 1 Corinthians 13:11, "When I was a child, I used to speak like a child, think like a child, reason like a child; when I became a man, I did away with childish things." Again, in Philippians 2:12–13, immediately after instructing us to take on Christ's self-sacrificing humility, Paul states, "Work out your own salvation with fear and trembling; for it is God who is at work in you both to will and to work for his good pleasure" (ch. π). God gave us the faculties of reason for a reason (ch. i'). It is God who performs the saving work (John 3; Romans 3–6; Ephesians 2; Hebrews 9–10), but it is our responsibility to seek God's desire and live it out.

Jesus told his disciples to "be as shrewd as serpents and as innocent as doves" in Matthew 10:16. Paul repeatedly appeals to his audiences' logical, intellectual faculties.

> Let no one deceive himself. If anyone among you thinks he is wise in this age, let him become a fool so that he can become wise. (1 Corinthians 3:18)

> But I fear that, as the serpent deceived Eve by his cunning, your minds may be seduced from a sincere and pure devotion to Christ. (2 Corinthians 11:3)

> I am amazed that you are so quickly turning away from him who called you by the grace of Christ and are turning to a different gospel—not that there is another gospel, but there are some who are troubling you and want to distort the gospel of Christ. (Galatians 1:6–7)

> Let no one deceive you with empty arguments, for God's wrath is coming on the disobedient because of these things. Therefore, do not become their partners. (Ephesians 5:6–7)

> I am saying this so that no one will deceive you with arguments that sound reasonable. (Colossians 2:4)

Consider the writing of John, one of Jesus's disciples. At the end of the book that bears his name, he states, "And there are also many other things that Jesus did, which, if every one of them were written down, I suppose not even the world itself could contain the books that would be written" (21:25). But prior to that, at the end of the previous chapter, he says, "Jesus performed many other signs in the presence of his disciples that are not written in this book. But these are written so that you may believe that Jesus is the

Messiah, the Son of God, and that by believing you may have life in his name" (20:30–31). John says that he couldn't have written down everything that Jesus did, but he has intentionally chosen certain events to record for the express purpose of serving as evidence to the reader, who after careful consideration may believe that Jesus is the Christ. John didn't come to his conclusion on a whim or based on emotion; he documented the substance of his claim.

From these three examples of key New Testament figures, it appears that Christianity requires high-functioning reasoning skills. So why do Christians have a reputation for mindless obedience? The author is certainly not an advocate of believing without evidence or explanation. Even when Jesus responds to Thomas, his disciple, in John 20:29, by saying, "Because you have seen me, you have believed. Blessed are those who have not seen and yet believe," he is not criticizing Thomas for lack of blind faith. This passage has been misconstrued perhaps by religious leaders in order to force blind obedience and by nonreligious individuals to accuse Christians of that blind obedience.

In reality, Thomas had an abundance of evidence to support belief that Jesus had risen from the dead. Jesus had performed a variety of other miracles defying nature, including raising Lazarus from the dead (John 11). Jesus had predicted that he would be killed by the Jewish leadership and would rise again on the third day (Luke 9:22). Mary Magdalene had testified that she had seen Jesus (John 20:18), and then the other disciples had shared their experience with Thomas (John 20:25a). Jesus did not criticize Thomas because he hadn't believed without evidence. Jesus's comment to Thomas was that he hadn't believed in spite of the evidence (ch. i^i). Thomas rejected historic events he had witnessed and personal testimony from his friends until he could see Jesus face to face. And even in that, Jesus lovingly met Thomas where he was. Responding to Thomas's proclamation: "If I don't see the mark of the nails in his hands, put my finger into the mark of the nails, and put my hand into his side, I will never believe" (John 20:25b), Jesus said, "Put your finger here and look

How Can a Rational God Allow Irrational Numbers?

at my hands. Reach out your hand and put it into my side. Don't be faithless, but believe" (John 20:27).

According to a book by Marco Livio, the golden ratio (φ) has been studied by some of the greatest mathematical minds in history. "But the fascination with the Golden Ratio is not confined just to mathematicians. Biologists, artists, musicians, historians, architects, psychologists, and even mystics have pondered and debated the basis of its ubiquity and appeal. In fact, it is probably fair to say that the Golden Ratio has inspired thinkers of all disciplines like no other number in the history of mathematics."[1] The golden ratio represents the comparison of two features that are in perfect aesthetic proportion. It is also an irrational number, as we saw in chapter $\sqrt{2}$. In a numeric sense, two line segments are said to be in the golden ratio if the ratio of the length of the greater to the length of the lesser is the same as the ratio of the sum of the lengths to the length of the greater. In other words, if a and b are real numbers, where $a>b$, then a and b are in the golden ratio if $a/b=(a+b)/a$. Fibonacci numbers can be used to approximate the value of φ; the estimation is more accurate for higher values of Fibonacci numbers, wherein element n is divided by element $n-1$; $φ≈F(n)/F(n-1)$.

Life is really about balance. Society questions what a healthy work-life balance would be; perhaps a 40-hour workweek is unrealistic. Should we work fewer hours? Many advocates of this shorter workweek do not seem to understand word problems, because they believe the compensation associated with fewer hours should be the same as with 40 hours. If we considered the golden ratio, perhaps we should be working 64 hours per week, since $104/64≈168/104≈1.618$. On the other hand, if we assume people sleep 8 hours per night (56 hours per week), then we have only 112 waking hours in a week. In which case, the golden ratio would be found in a workweek of approximately 43 hours, since $69/43≈112/69≈1.618$. Perhaps the 40-hour workweek makes sense after all.

What about balance in our insatiable thirst for knowledge and our ability to accept incomplete information in making decisions?

1. Livio, *Golden Ratio*, 6.

Y AND YO: PROBLEM-SOLVING

How do we find the golden ratio of critical mass of knowledge to total knowledge? Paul tells us "For now, we see in a mirror dimly, but then face to face; now I know in part, but then I will know fully" (1 Corinthians 13:12). The author agrees with Paul that we as humans will never have a perfect knowledge in this life. In professional circles, we talk about getting a 90-percent answer or an 80-percent answer to convey some confidence that we have sufficient information to make a reasonable judgment call. Perhaps the golden ratio suggests that 61.8 percent is sufficient, since another of the phenomenal properties of the golden ratio is that $1/\varphi = \varphi - 1$ or $(1/1.618) = (1.618 - 1) \approx 0.618$.

Regardless of one's desire to study and research, he/she must make judgment calls every day based on incomplete facts. Sometimes we have more than enough information, and other times we lament the paucity of data points. In either circumstance, however, we must make choices based on whatever we have. Many may choose to blindly believe the word of others without contest or question. However, as Socrates said in a different context, "The unexamined life is not worth living."[2] The author is not interested in unexamined, untested, or house-of-cards religious beliefs but rather in concrete, life-changing faith that is well thought out and explained.

With that understanding, the next few chapters describe an interrogation of the Scriptures on the what, how, and why of Christianity. For those intellectuals who grew up being told that they must *just believe*, the author hopes that the following irrational chapters, which constitute the author's process of working out his own salvation (Philippians 2:12), will be helpful in elaborating the bigger picture of Christianity. The author also encourages the reader to not accept at face value the interpretations presented herein but rather to take up the Scriptures for himself/herself to scrutinize and examine them personally as the Bereans did (Acts 17:10–15).

---

2. Kraut et al., "Socrates," s.v. "Plato," para. 6.

Chapter e

$de^x/dx=e^x$: Transcendent Purpose

TRANSCENDENTAL NUMBERS ARE COMPLEX numbers that are not solutions to polynomial equations with rational coefficients and integer exponents. While there are believed to be infinitely many transcendental numbers, it is difficult to prove that a number is transcendental. Euler's number (e) is an example of a transcendental number.[1] All transcendental numbers are irrational, but they are even more unique since they also transcend the mere mortal, rational polynomial world.

Euler's number is one of the most phenomenal numbers in the set of all complex numbers. The fact that the derivative of e^x with respect to x is e^x, that is $de^x/dx=e^x$, is mind blowing. The rate of change of the function at a specific point x is the value of the function at that specific point. Pause and take that in. The very existence of this number is almost sufficient, in and of itself, to suggest an intelligent design to the universe. In the upcoming chapters numbered as multiples of e, we will consider *what*-type questions.

Just as e is an element of the set of all complex numbers, we are called as individuals to be elements of the world's population. Just as e is unique in its transcendence or overcoming of the mere mortal, rational polynomial world, God has called each of us to

---

1. Müller, "Transcendental Numbers."

DEX/DX=EX: TRANSCENDENT PURPOSE

a transcendent purpose. This chapter will focus on God's call to humanity to that transcendent purpose.

As humans, we seek meaning in the universe. That is the best explanation for our endless pursuit of knowledge. Throughout history, cultures have looked to the heavens and the stars for guidance and navigation in both a physical sense and an existential sense. The numerous horoscope and zodiac columns found online and in mobile apps are clear evidence of this. As humans, we want to transcend our own natural limits. Enlightenment, both the religious meaning and the rational movement, attempts to achieve this transcendence we seek as individuals and as a whole. Unfortunately, enlightenment in either of these avenues focuses on the individual's relationship to the greater entity or to the universe or to knowledge. As an intellectual community, since we have rejected faith as an explanation for the universe, we have also failed to see that the Creator called us to a transcendent purpose that originates with the Creator rather than with us. The reader may argue semantics; however, our transcendent purpose is not centered on our relationship to the Creator but rather the Creator's relationship to us.

We don't have to look very deep into the history of humanity to find the first example of when we demonstrate our pursuit of purpose, knowledge, and meaning. In Genesis 3, Adam and Eve rejected God's instruction to not eat of the tree of knowledge of good and evil, because the serpent explained that it would grant them godlike understanding. In fact, the serpent questioned God's motives for forbidding them from eating the fruit (3:4–5). This first sin stemmed from pride, our desire to be like God. We want to know. And then we want others to know that we know. It wasn't sufficient for Eve to eat the fruit; she wanted Adam to join her (3:6).

In Genesis 11, humans demonstrated once again our arrogance. "And they said, 'Come, let us build ourselves a city and a tower with its top in the sky. Let us make a name for ourselves; otherwise, we will be scattered throughout the earth'" (v. 4). We are still seeking to make a name for ourselves today. How many hospital or educational buildings are named for philanthropists or

How Can a Rational God Allow Irrational Numbers?

businesses are named for founders? We seek to be remembered; we want to transcend our own lifetimes. The book of Ecclesiastes tells us that God "has placed eternity in [our] hearts" (3:11). But he also gave us purpose and meaning. He gave us the transcendent cause not to exalt our own names but rather to exalt his.

God called Abram (later Abraham) to leave his comfortable life and go wander to a new place. While we desire to make our own names great by the achievements we accomplish, God promised to make Abram's name great on the condition that Abram be obedient. Abram was called to a transcendent future that God originated; "all the peoples on earth will be blessed through you" (Genesis 12:3).

God called Moses from his ineptitude to challenge Pharaoh and lead the people of Israel to the promised land. "Who am I that I should go to Pharaoh and that I should bring the Israelites out of Egypt?" (Exodus 3:11). God called Gideon from his insignificance to deliver the people of Israel from the Midianites. "Look, my family is the weakest in Manasseh, and I am the youngest in my father's family" (Judges 6:15). Clearly, God was not looking for, nor did he need, a *somebody*.

When God told the prophet Samuel to anoint the second king of Israel because the first king, Saul, had been disobedient and rebellious, Samuel went as directed to Bethlehem (1 Samuel 16:1–4). There he met Jesse, whom he directed with his sons to consecrate themselves and accompany him to the sacrifice. When the family arrived, Samuel saw Eliab, Jesse's oldest son, and his physical appearance was aesthetically pleasing. Samuel thought that Eliab was to be king (16:5–6). God spoke to Samuel though and said, "Do not look at his appearance or his stature because I have rejected him. Humans do not see what the Lord sees, for humans see what is visible, but the Lord sees the heart" (16:7). Samuel looked on as Jesse paraded seven of his sons, and God chose none of them (16:8–10). Samuel asked, "Are these all the sons you have?" (16:11). Then Jesse remembered his youngest son, the one who was keeping the sheep. David was not even considered worthy of calling to join the family for the sacrifice. When

DEX/DX=EX: Transcendent Purpose

he arrived, God told Samuel to anoint him to be the next king of Israel. David was called from relative obscurity to be the king of Israel (16:12). David was not going to make his own name great; God was going to make David's name great.

Later, when this same David promised King Saul that he would fight the giant Goliath, King Saul responded with doubt: "You can't go fight this Philistine. You are just a youth, and he's been a warrior since he was young" (1 Samuel 17:33). But David assured Saul, "Your servant has been tending his father's sheep. Whenever a lion or a bear came and carried off a lamb from the flock, I went after it, struck it down, and rescued the lamb from its mouth. If it reared up against me, I would grab it by its fur, strike it down, and kill it. Your servant has killed lions and bears; this uncircumcised Philistine will be like one of them, for he has defied the armies of God. . . . The Lord who rescued me from the paw of the lion and the paw of the bear will rescue me from the hand of the Philistine" (17:34–37). Here we see an exhibition of David's faith (ch. i'). Notice again that his faith is not blind and uneducated; it is firmly founded on historical experience when God had delivered his enemies into his hands. David was a nobody, but he had been faithful in service to his father, and it was in the pastures of his life that David had developed the heart of God's chosen leader to be king. God wasn't looking for someone who could be king; he was looking for someone who would be faithful and obedient. Jesus tells us, "Whoever is faithful in very little is also faithful in much, and whoever is unrighteous in very little is also unrighteous in much. So if you have not been faithful with worldly wealth, who will trust you with what is genuine? And if you have not been faithful with what belongs to someone else, who will give you what is your own?" (Luke 16:10–12).

Esther was exalted to queen from humble servitude. Her willingness to be bold in the face of fear rescued her entire nation from destruction. In appealing for her to trust God to deliver her from certain death for entering the king's presence unbidden, her cousin Mordecai challenged her, "Who knows, perhaps you have come to your royal position for such a time as this" (Esther 4:14).

How Can a Rational God Allow Irrational Numbers?

God promoted Esther, the pawn, to queen, but he had not done so just to exalt her name. He had a purpose for her to serve in the position where he placed her. God calls us to a transcendent purpose through Jesus Christ, and we can be confident that the circumstances and challenges we face are meant to teach us to rely on him.

Some of these individuals identified themselves as weak or incompetent. Some of them were labeled by others with these attributes. But God, seeing the heart of humans, knows us for who we are; he knows our true identities. When we find our identities in God, we are able to be who we were truly meant to be (ch. $e^{i\pi}$).

So, what is this grand purpose to which God has called us? There are several instances where the Bible gives us a broad sense of God's design for humanity. In Leviticus 11:44, God tells the Israelites to "consecrate yourselves and be holy because I am holy." This command is given in relation to dietary habits, and while in the New Testament God relaxes the dietary restrictions, the general principle of being holy is not rescinded. In fact, Jesus tells us, "Be perfect, therefore, as your heavenly Father is perfect" (Matthew 5:48). Paul urges us to "live worthy of the calling we have received" (Ephesians 4:1). Peter tells us, "As obedient children, do not be conformed to the desires of your former ignorance. But as the one who called you is holy, you also are to be holy in all your conduct" (1 Peter 1:14–15).

In Mark 12, Jesus is asked to simplify God's desire for humans. "'Which command is the most important of all?' Jesus answered, 'The most important is Listen, O Israel! The Lord our God is one. Love the Lord your God with all your heart, with all your soul, with all your mind, and with all your strength. The second is, Love your neighbor as yourself. There is no other command greater than these'" (Mark 12:28–31). While these two commandments simplify the depth and breadth of the Old Testament law, they do not provide the specific particulars of everyday behavior. Loving God requires us to know God and to value the things he values. Since God is the eternal Being, he is transcendent. When we submit ourselves to him and emulate his character, we are fulfilling the transcendent purpose to which he has called us. This

DEX/DX=EX: TRANSCENDENT PURPOSE

is why the general principle of being holy as God is holy is not nullified even though the particulars of dietary regulation are made obsolete. Loving other people is actually an extension of loving God, because when we truly love God by knowing him and emulating his character, we value what he values. God esteems humanity and each individual person as valuable (ch. $\sqrt{2}$). Christ died for humanity because he loved each of us. From these two general principles of loving God and loving people, we understand why God tells us to care for the helpless, the weak, the widow, the orphan, the foreigner, etc. Since God values human life, we should care for others.

In Genesis 1:26–31, we are told that God created humans in his image. This communicates to us that God designed us to bear and reflect his image. We were also created for family and community, "Then the Lord God said, 'It is not good for the man to be alone. I will make him a helper corresponding to him'" (Genesis 2:18). Our function in family and community is further discussed in chapter 7[3], but at this point let it suffice the reader to understand that if we are to bear and reflect God's image, there should also be an audience to observe God's handiwork in our lives.

Perhaps the clearest explanation of our transcendent purpose is presented in Paul's words in Ephesians 2:6–7, "He also raised us up with him [Christ] and seated us with him in the heavens in Christ Jesus, so that in the coming ages he might display the immeasurable riches of his grace through his kindness to us in Christ Jesus." God's transcendent purpose for us does not originate with us but with him. Our purpose is to glorify him as he has blessed us. "For we are his workmanship, created in Christ Jesus for good works, which God prepared ahead of time for us to do" (2:10). God has not called us to a purpose that focuses on us and our ephemerality; no, he has called us to his transcendent purpose in eternity.

Chapter π

y: The Mystery

IRRATIONAL NUMBERS WERE NOT discovered until sometime around 500 BC,[1] but, interestingly, early estimates of π date back four thousand years.[2] In 1 Kings 7:23, when Solomon had the temple built (perhaps 1000 BC), π was estimated to be three since a specific pillar is described as having a circumference equal to three times the diameter. This seems uneducated and incorrect, imprecise at the very least. Certainly, if there were an all-knowing God who created the universe, who also wrote the Bible, he would know the true value of π. The author will grant the reader that such a God would know the precise value of the infinite decimal places of π but reminds the reader that irrational numbers might have only just been in the process of being identified and defined by humans in certain parts of the world at that point, and Israel does not seem to have been among those.

The exact value of π is a mystery; it escapes us. Even though humans have derived over six billion digits of π, we cannot know its value perfectly.[3] This is both frustrating and fascinating. Think for a moment; there are levels of precision that we cannot comprehend.

1. Centrone, "Pythagoras."
2. Ye and The Conversation, "Long Search."
3. Robinson, "What Is Pi?"

Y: THE MYSTERY

There are dimensions we are unable to fathom, wavelengths we cannot perceive, and aspects of the universe we have yet to discover. There is a sense of wonder or mystery to these concepts (see ch. *i*, the known unknown and the unknown unknown). Equally mysterious are the concepts of faith, salvation, and our obedience to God. In algebra, we are often asked to determine the value of (or the nature of) the dependent variable, y. A similar question arises here. Why (y)? The upcoming chapters, numbered as multiples of π, will focus on *why*-type questions. What is this mystery of the gospel, and why is it significant, and what should my response be to it?

There's a whole genre of entertainment (books, movies, apps, live-action events, etc.) dedicated to solving mysteries. We describe certain complex events as mysterious such as the navigational troubles surrounding the Bermuda Triangle. We describe uncertainty in historical events as mysterious, such as the final whereabouts of Amelia Earhart. What is our cultural draw to mystery? Why do "inquiring minds want to know"?

The author of Proverbs instructed us repeatedly to seek wisdom and knowledge (Proverbs 1:1–7). It is somewhat paradoxical that, as wise as we are as intellectuals, this mystery of a transcendent purpose, which we have sought through various forms of enlightenment and self-actualization, has escaped us. In the book of Daniel, King Nebuchadnezzar could not remember a troubling dream. When his wise men and mediums could not recount and interpret it, he became disillusioned with them and ordered them all to be executed. But Daniel sought God's counsel. Before he returned to the king with the dream's content and its interpretation, we find Daniel's personal praise of God:

> May the name of God be praised forever and ever, for wisdom and power belong to him. He changes the times and the seasons; he removes kings and establishes kings. He gives wisdom to the wise and knowledge to those who have understanding. He reveals the deep and hidden things; he knows what is in the darkness, and light dwells with him. I offer thanks and praise to you, God

of my fathers, because you have given me wisdom and power. And now you have let me know what we asked of you, for you have let us know the king's mystery. (Daniel 2:20–23)

Daniel humbly sought God's wisdom to reveal the mystery. When he was presented to the king,

> the king said in reply to Daniel . . . "Are you able to tell me the dream I had and its interpretation?" Daniel answered the king: "No wise man, medium, magician, or diviner is able to make known to the king the mystery he asked about. But there is a God in heaven who reveals mysteries, and he has let King Nebuchadnezzar know what will happen in the last days." (2:26–28)

If God is indeed the Revealer of mysteries and has told us how to investigate mysteries, then it would follow that a learned person would discover this mysterious transcendent purpose. It is a word problem, and we have the ability to interrogate the Creator (ch. φ).

We have already discussed the origins of the universe (ch. o) and the current plight of humanity in the world (ch. 1). We have discussed the irrationality of God sacrificing himself for us, his enemies (ch. $\sqrt{2}$). While we have explored these mysteries in some depth, let us consider now what Paul calls the "mystery of the Christ" (Ephesians 3:4).

Why did Jesus speak in parables? When the disciples asked this very question in the book of Matthew, Jesus answered, "Because the secrets of the kingdom of heaven have been given for you to know, but it has not been given to them. . . . That is why I speak to them in parables, because looking they do not see, and hearing they do not listen or understand." Jesus went on to quote the prophet Isaiah: "For this people's heart has grown callous; their ears are hard of hearing, and they have shut their eyes; otherwise they might see with their eyes, and hear with their ears, and understand with their hearts and turn back—and I would heal them" (Matthew 13:11–15; quoting Isaiah 6:9–10). The mystery of God's design for the universe is not a secret because God has not revealed it but rather because humanity has rejected it.

Y: THE MYSTERY

Paul tells us that he is sharing his insight into this mystery. "This was not made known to people in other generations as it is now revealed to his holy apostles and prophets by the Spirit" (Ephesians 3:5). If everyone already knew a secret, it wouldn't be much of a mystery. The Old Testament law was given to the Jews at the hand of Moses. The Jews believed that salvation would come in the form of a Messiah for their people alone. After all, the eternal throne was promised to David (2 Samuel 7:16). But they misunderstood God's comprehensive and overwhelming love. As God promised Abram, "all the peoples on earth will be blessed through you" (Genesis 12:3). This is a mystery to the Israelites as well as the Gentiles because the Israelites were not being faithful to share this good news of God's design and purpose for humans. In 1 Corinthians 4:1-2, Paul tells us that he is a servant of Christ and manager of the mysteries of God; furthermore, managers must be found faithful.

Paul continues writing to the Ephesians, "The Gentiles are coheirs, members of the same body, and partners in the promise in Christ Jesus through the gospel. I was made a servant of this gospel by the gift of God's grace that was given to me by the working of his power" (Ephesians 3:6-7). Paul recognized his position as a servant to share the gospel with those who the Israelites had considered unworthy of the Messiah. "This grace was given to me . . . to proclaim to the Gentiles the incalculable riches of Christ, and to shed light for all about the administration of the mystery hidden for ages in God who created all things" (3:8-9). This inclusive nature of God's love should not have been a mystery, but it was, and Paul was elected as the servant to administer the gospel to the outcasts (everyone outside the Jewish race). However, we should not fail to notice the mystery of the infinite nature of God's unfailing love that Paul denotes as the "incalculable riches of Christ."

Paul finally tells us more specifically what the mystery of Christ is in the next verses: "This is so that God's multifaceted wisdom may now be made known through the church to the rulers and authorities in the heavens. This is according to his eternal purpose accomplished in Christ Jesus our Lord" (Ephesians

How Can a Rational God Allow Irrational Numbers?

3:10–11). The mystery, is in fact, the eternal purpose that God accomplished through the life, death, and resurrection (the gospel) of Jesus Christ. In 1 Timothy 3:16, Paul verifies this when he says, "And most certainly, the mystery of godliness is great: He [Christ] was manifested in the flesh, vindicated in the Spirit, seen by angels, preached among the nations, believed on in the world, taken up in glory." Jesus fulfilled the purpose God laid out for him and was taken up in glory.

In his Letter to the Colossians, Paul makes it clear that God desired to reveal his mysteries to all, not to preserve them for a sacred few:

> Now I rejoice in my sufferings for you, and I am completing in my flesh what is lacking in Christ's afflictions for his body, that is, the church. I have become its servant, according to God's commission that was given to me for you, to make the word of God fully known, the mystery hidden for ages and generations but now revealed to his saints. God wanted to make known among the Gentiles the glorious wealth of this mystery, which is Christ in you, the hope of glory. We proclaim him, warning and teaching everyone with all wisdom, so that we may present everyone mature in Christ. (1:24–28)

God's goal was to make himself known, to bestow "the incalculable riches" or "the glorious wealth of this mystery" on us. That mystery is Christ in you; he is the hope of glory.

At the end of the previous chapter, we identified God's transcendent purpose for us to showcase the riches of his grace that he bestowed on us in Christ. The mystery of the gospel is found in what we have already identified, that the Creator sacrificed himself for the creation in order to display his magnificent and eternal benevolence. In chapter $\sqrt{2}$, we denoted the sacrifice of Christ for his enemies as irrational; this certainly qualifies it as a mystery. When we humbly receive this sacrifice and then devote our lives to sharing this redemptive message with others, God's magnificence is on display in us through our godliness and Christlikeness, through our living out the Christian life.

y: The Mystery

Paul continues on to pray for the Ephesians as he says,

> For this reason I kneel before the Father from whom every family in heaven and on earth is named. I pray that he may grant you, according to the riches of his glory, to be strengthened with power in your inner being through his Spirit, and that Christ may dwell in your hearts through faith. I pray that you, being rooted and firmly established in love, may be able to comprehend with all the saints what is the length and width, height and depth of God's love, and to know Christ's love that surpasses knowledge, so that you may be filled with all the fullness of God. (Ephesians 3:14-19)

Here Paul clears up any doubts we had as to whether God's love were boundless. Is there any greater mystery than to understand that which surpasses knowledge? How can we be filled with all the fullness of God? These are additional mysteries of Christ and his incalculable riches. Paul finishes this chapter by stating,

> Now to him who is able to do above and beyond all that we ask or think according to the power that works in us—to him be glory in the church and in Christ Jesus to all generations, forever and ever. Amen. (3:20-21)

As we have seen, Paul tells the Ephesians that we "are saved by grace through faith," not by our works (Ephesians 2); however, Paul tells the Philippians to "work out [their] own salvation" (Philippians 2:12). He is clearly not telling the Philippians to save themselves. He is instructing them (and us) to delve into the mysteries of salvation, to grow beyond the milk and to grow into mature believers (ch. 2π). Remember from chapter φ, Paul tells us in 1 Corinthians 13:11, "When I was a child, I spoke like a child, I thought like a child, I reasoned like a child. When I became a man, I put aside childish things." He tells Timothy to diligently study, so that he may demonstrate himself unashamed before God, correctly teaching the truth (2 Timothy 2:15).

The mystery of the gospel is like Russian nesting dolls. Once we think we have resolved the mystery, the origin of the universe, we find that the Creator has sacrificed himself for the creation.

Then we find a new mystery that God's love is boundless and inclusive, though it seemed limited and exclusive. Finally, we see that there are deeper mysteries still to be searched out in God's infinite power and provision. If we are to discover our transcendent purpose, we must seek the mysteries that God is revealing to us and live them out. Notice that Daniel and Paul both ascribe praise, honor, and glory to God. This is our ultimate transcendent purpose, to display God's eternal, unfailing, incalculable love for his creation.

As Paul told the Colossians and the Laodiceans, the author offers this prayer for the reader: "I want [your heart] to be encouraged and joined together in love, so that [you] may have all the riches of complete understanding and have the knowledge of God's mystery—Christ. In him are all the treasures of wisdom and knowledge" (Colossians 2:2–3).

Chapter 2e

d/dx: Spiritual Growth

THE AUTHOR HAS OFTEN lamented the way mathematics textbooks were written in his school days. In any math class, one pressing question is always posed by nearly every student in the classroom: "Why do I need to know this?" or "When will I ever use this again?" The author believes this question could be avoided altogether by beginning new chapters with the answer to a *what* question instead. For instance, consider this scenario: On the first day of Calculus I class, we need to introduce the concepts of limits and derivatives.

Have you ever been on an open highway and tried to drive your car as fast as it could possibly go? If anyone confesses, ask "What happened?" (If no one will own up to having tried, describe the process.) What happens is your car initially accelerates quickly and then the rate of acceleration decreases, the vehicle is gaining speed at a slower and slower rate, and it reaches maximum speed—it cannot go any faster; it doesn't quite reach the maximum theoretical rate (which may or may not be on the speedometer), but that is the upper *limit* (draw a representative graph for visual effect). What does the speedometer measure? It measures instantaneous velocity; it is the instantaneous rate of change or the *derivative* of your position function. What is acceleration? It is the rate of change of your velocity, so it is also a derivative. Does anyone in the room have a job? How much do you get paid per hour? That pay rate is actually a derivative of your wealth function. Now let's delve into limits and derivatives to understand them better.

While the author is an avid lover of the *why* question, we do ourselves a disservice in asking *why* before we answer *what*. *Why* can be properly asked, investigated, and understood only when we fully grasp *what*. Our students would be better served to receive knowledge before asking why they should receive it. However, since it seems to be human nature to desire a minimalist's knowledge, educators would do well to circumvent a student's natural tendency to reject new material until they know why they should be interested by demonstrating the usefulness of the material first. Following this theme, this chapter is focused on spiritual growth, or the derivative of spiritual maturity. We will look at the *why* aspect (that is, what spiritual maturity is and how to achieve it) in the next chapter.

Once you make the decision to follow Christ, a new dimension is added to your life. You have been brought to spiritual life. In this new dimension, our goal is to grow spiritually. Think of an increasing function starting at the origin. Let this function define our spiritual growth (the rate at which we are gaining spiritual maturity); it could take on many shapes. It can grow linearly, logarithmically, hyperbolically, exponentially, etc. Does it grow without bound or reach an upper limit? The nature of this function, how you grow spiritually, depends on which steps you take to achieve growth.

What does it mean to grow? When a child is born, he/she feeds on milk to grow or mature physically. The rate at which this person grows is dependent on the diet he/she intakes. Naturally, over time, the child also begins to mature, as his/her brain and body develop. There are various aspects of maturity, but how maturity is gained and how fast maturity is gained depend on innate characteristics, nutrition, environment, experiences, etc. Once a person approaches the cross and accepts God's gift of salvation and takes on God's transcendent purpose, this individual begins to grow spiritually.

Jesus calls the initiation of this process being "born again" (John 3:3) and "passing from death to life" (John 5:24). In John

11, we are told the story of the resurrection of Lazarus, Jesus's friend. This physical event demonstrates many of the spiritual steps necessary for a new believer. First, Jesus commands the stone be removed from the grave (v. 39). Believers must remove barriers to the gospel from the path of lost people. Second, it is the responsibility of Christians to engage the lost world (vv. 39–41). Believers must be obedient to spend time with lost people, unlike Martha, who complained that the dead man stank. Third, salvation is always for God's glory (vv. 41–42). It is not our eloquence that calls the dead to life; Jesus recognizes God as the supreme authority. Fourth, Jesus calls the dead man to life (v. 43). Jesus is the only one with the power to grant life to the dead; we are merely responsible for our obedience. Fifth, it is the responsibility of mature believers to help the new believer learn to walk in freedom from bondage to sin (v. 44). Jesus commands those standing around, "Unwrap him and let him go." This final point is the part we most often fail to fulfill as believers; it entails helping the newborn believer to grow spiritually (see ch. 7[3]).

Paul gives us some practical advice for what it looks like to grow spiritually. He begins with instructions on what not to do in Ephesians 4: "Therefore, I say this and testify in the Lord: You should no longer live as the Gentiles live, in the futility of their thoughts. They are darkened in their understanding, excluded from the life of God, because of the ignorance that is in them and because of the hardness of their hearts. They became callous and gave themselves over to promiscuity for the practice of every kind of impurity with a desire for more and more" (vv. 17–19). Paul tells us first that we must leave these old behaviors behind. After all, unbelievers have rejected the mystery of the gospel, and we have accepted it through salvation and are on a journey to fulfill our transcendent purpose. If we as *mature* Christians are still living worldly lives, we cannot possibly help new believers grow in Christ.

Paul continues, "But that is not how you came to know Christ, assuming you heard about him and were taught by him, as the truth is in Jesus, to take off your former way of life, the old self that is corrupted by deceitful desires, to be renewed in the spirit of your minds, and to put on the new self, the one created according to God's likeness in righteousness and purity of the truth" (vv. 20–24). Paul tells us to take off the old, corrupt self in these sinful behaviors, and put on the new self that God has granted us in Christ Jesus. Just as Jesus commands that Lazarus's grave clothes be removed, Paul instructs us to get rid of our old habits. The new believer needs help identifying sinful behaviors to be removed and new godly behaviors that he/she should be putting on.

Paul then proceeds to advise us by listing some examples of both behaviors:

> Therefore, putting away lying, speak the truth, each one to his neighbor, because we are members of one another. Be angry and do not sin. Don't let the sun go down on your anger, and don't give the devil an opportunity. Let the thief no longer steal. Instead, he is to do honest work with his own hands, so that he has something to share with anyone in need. No foul language should come from your mouth, but only what is good for building up someone in need, so that it gives grace to those who hear. And don't grieve God's Holy Spirit. You were sealed by him for the day of redemption. Let all bitterness, anger and wrath, shouting and slander be removed from you, along with all malice. And be kind and compassionate to one another, forgiving one another, just as God also forgave you in Christ. (vv. 25–32)

The author of Hebrews indicts his audience, "Although by this time you ought to be teachers, you need someone to teach you the basic principles of God's revelation again. You need milk, not solid food. Now everyone who lives on milk is inexperienced with the message about righteousness, because he is an infant. But solid food is for the mature—for those whose senses have been trained to distinguish between good and evil" (Hebrews 5:12–14). In general, we do not like to be called babies, but that is precisely what this author is calling the believers in the audience. It is our responsibility to grow into spiritual maturity, which we will address in the next chapter. This means that we must move beyond the basic principles of God's revelation and take on the mysteries of the gospel. Spiritual growth involves training our senses to discern good and evil. This requires discipline.

Peter tells us, "[God]'s divine power has given us everything required for life and godliness through the knowledge of him who called us by his own glory and goodness. By these he has given us very great and precious promises, so that through them you may share in the divine nature, escaping the corruption that is in the world because of evil desire" (2 Peter 1:3-4). Here we find that, in Christ, we have been given all we need to live godly lives; God has granted us a divine nature, a nature distinct from our human nature, which is self-interested. When we enter into a relationship with Christ, we take on his character. This character must be developed; that is what spiritual growth means. We spend time studying who God is, and we begin to discipline ourselves to emulate him through the new, divine nature he has granted us; it is a process.

Peter continues to explain that while faith yields salvation, it is insufficient to attain spiritual maturity. He provides this advice to help us grow spiritually: "For this very reason, make every effort to supplement your faith with goodness, goodness with knowledge, knowledge with self-control, self-control with endurance, endurance with godliness, godliness with brotherly affection, and brotherly affection with love. For if you possess these qualities in increasing measure, they will keep you from being useless and unfruitful in the knowledge of our Lord Jesus Christ" (2 Peter 1:5-8). The author of Hebrews chided the readers for their inability to serve as teachers. Peter tells us that we can avoid being useless by growing spiritually in specific behavioral disciplines.

Paul tells Timothy, "Have nothing to do with silly myths. Rather, train yourself in godliness. For the training of the body has limited benefit, but godliness is beneficial in every way, since it holds promise for the present life and also the life to come" (1 Timothy 4:7-8). Even in the Christian faith, there are those who pursue foolishness instead of godliness. We must discipline ourselves to pursue godliness and abandon childishness and selfishness as we aspire to attain spiritual maturity.

Chapter 2π

∫x/dx: Spiritual Maturity

WHILE A DERIVATIVE TELLS us the rate at which a function is changing and what that change looks like, the integral of a function tells us more about why the change is occurring. Consider your existence on this planet. Every day, you get one day older; that is your physical body is aging a specific amount for each day that passes. But what does this aging process describe? The aging process is actually a derivative of another function, Age. We may write this as $d\text{Age}/dt$ = 1 day of Age per 1 day of t. But why is this equation significant? I have never heard anyone ask, "How are you aging?" Instead, people ask, "How old are you?" Some even become unrealistically optimistic and ask, "How young are you?"

Integration of the aging function yields Age; we could write $\int d\text{Age}/dt\ (t|t_0 \text{ to } t)$ = Age. This is the measurable quantity about which we inquire or lament. While physical age is a single-variable function of time, maturity is a multivariate function. It includes the independent variable of time as well as other variables that are codependent, such as experiences, environment, education, etc. This leads us to the author's favorite general equation that describes a person's state of being:

∫x/dx: Spiritual Maturity

$$\stackrel{\text{\Large\textlceil}}{\wedge}_t = \sum_{i=0}^{t-1} \left(\stackrel{\text{\Large\textlceil}}{\wedge}_i \right)$$

Think about the equation for a moment. What does the left-hand side represent? Now consider the right-hand side. Stickman at some point t is equal to the summation of Stickman over the period from 0 to t-1.

You are the sum of your yesterdays. If the author were ever to get a tattoo, this would be it. All the forces, decisions, and encounters that you have experienced (and not experienced) in the past make you who you are today. Aristotle, Descartes, and Locke would all likely approve. Disregarding supernatural forces, this equation is trivial; one can respond today based only on his/her experiences in the past. However, in the spiritual realm, this equation still holds true. The difference is that in the Christian's yesterdays, he/she encountered a supernatural force that changed his/her nature. At some point in the Christian's past, he/she encountered a new force in Jesus Christ that disrupted his/her natural maturity process and caused a jump in his/her growth function. Beyond that, it added a new dimension to his/her life; one to which he/she was oblivious until having met Jesus. He/she began maturing in the spiritual plane; he/she was born into this new life. This encounter with Jesus brought him/her to spiritual life. As 2 Corinthians 5:17 says, "If anyone is in Christ, he is a new creation; the old has passed away, and see, the new has come."

This equation was particularly meaningful to the woman caught in adultery. The reader may be familiar with the historical account found in John 8. The Pharisees had caught a woman in the very act of adultery and brought her to Jesus. (It is a mystery how the Pharisees could substantiate their claim since they did not bring a conspirator with her.) When they asked Jesus whether they should follow the law of Moses to stone her, Jesus ignored them until they pressed him a second time. He replied that whoever was without sin should cast the first stone. The Pharisees, feeling convicted of their own shortcomings, left one by one. "When Jesus stood up, he said to her, 'Woman, where are they? Has no one condemned you?' 'No one, Lord,' she answered. 'Neither do I condemn you,' said Jesus. 'Go, and from now on do not sin anymore'" (vv. 10–11).

Jesus could have stopped after saying "Neither do I condemn you." After all, he told Nicodemus that he had not come into the world to condemn the world (John 3:17). Why did he follow up with "Go, and from now on do not sin anymore"? Because he came to bring her life and salvation. Her life had been changed after having a personal encounter with Christ. This equation above is not a definition. We are not defined by our pasts; we are refined by our pasts! Every encounter you have had and every encounter you have not had are experiences that you build on. The fact that the author has tried mint chocolate chip ice cream in the past and did not like it enables him to make a new choice today. The fact is that this woman met Jesus, and that encounter changed her life; she walked away different.

In the previous chapter we talked about the what and how of spiritual growth; now we are discussing the why, let us consider the impetus for spiritual growth: spiritual maturity. In general, the goal of the individual is to become a contributing member of society as an adult. This is a simple, working definition of the word "mature." There are many facets to this quality, such as being patient and kind, not screaming at your neighbor for some insignificant trespass, being responsible for your own actions, etc.

As a believer, the primary task is to train one's senses to discern good and evil (Hebrews 5:13–14) so one can behave like Christ. This is the definition of spiritual maturity. Spiritual maturity cannot be measured by the absence of profanity from your vocabulary, the lack of alcohol in your consumptive diet, or abstention from adulterous behaviors. While spiritual growth involves working on some of these aspects as identified in the previous chapter, spiritual maturity is the result of having trained our

∫x/dx: Spiritual Maturity

senses. Spiritual maturity is not a measure of what is missing from one's life; it is a measure of how Christlike one is.

Why should we grow spiritually? To be spiritually mature. What is the purpose of spiritual maturity? As the author of Hebrews stated, the spiritually mature can be teachers to guide others (see ch. 7³). On a more fundamental level, spiritual maturity is a reward in and of itself. Paul tells us, "Now the works of the flesh are obvious: sexual immorality, moral impurity, promiscuity, idolatry, sorcery, hatreds, strife, jealousy, outbursts of anger, selfish ambitions, dissensions, factions, envy, drunkenness, carousing, and anything similar" (Galatians 5:19–21). Avoiding these character traits requires discipline because they are natural tendencies of humanity, but consider the benefits of a life that is free from these maladies.

It is not sufficient to know the negative characteristics that we should clearly avoid; we must also know how we should be living. Paul continues, "But the fruit of the Spirit is love, joy, peace, patience, kindness, goodness, faithfulness, gentleness, and self-control" (Galatians 5:22–23). These are the attributes that we should exhibit in our lives as spiritually mature Christians; these characteristics reflect the heart of God. If the reader were to describe the primary character of Christianity based on the Christians in his/her circle, would any of these make the top ten traits identified? If these are not the characteristics readily identified, then most likely the Christian friends and acquaintances in question are not behaving like mature believers. Paul tells us, "Now those who belong to Christ Jesus have crucified the flesh with its passions and desires. If we live by the Spirit, let us also keep in step with the Spirit. Let us not become conceited, provoking one another, envying one another" (5:24–26).

Remember that we were created in Christ Jesus for a transcendent purpose. "He also raised us up with him [Jesus] in the heavens in Christ Jesus, so that in the coming ages he might display the immeasurable riches of his grace through his kindness to us in Christ Jesus" (Ephesians 2:6–7). We are made, then, to imitate his character in performing good works. "For we are his workmanship, created in Christ Jesus for good works, which God prepared ahead of time for us to do" (2:10). This is spiritual maturity: to be

Christlike. When we walk in those good works emulating Christ's character, God is glorified.

We must, as the author of Hebrews says, "lay aside every hindrance and the sin that so easily ensnares us. Let us run with endurance the race that lies before us, keeping our eyes on Jesus, the source and perfecter of our faith. For the joy that lay before him, he endured the cross" (Hebrews 12:1-2). Jesus, our supreme example of spiritual maturity, demonstrates faith through his faithfulness. The author tells us to lay aside the distractions the world presents to us so that we can run the race with endurance. Beginners do not run the race with the endurance described here; only those who have trained run this way. Immature believers are busy devising and articulating arguments about silly myths instead of growing spiritually through the discipline of training themselves in godliness (1 Timothy 4:7-8).

Paul commands the Ephesians to "live worthy of the calling [they] have received, with all humility and gentleness, with patience, bearing with one another in love, making every effort to keep the unity of the Spirit through the bond of peace" (Ephesians 4:1-3). Paul's instructions challenge us to live in spiritual maturity. Living in humility, patience, unity, and peace do not come accidentally; these take a significant amount of intentional effort and discipline. Spiritual growth leads to spiritual maturity, which manifests itself as godliness. To the Philippians, Paul says, "I have learned to be content in whatever circumstances I find myself. I know both how to make do with little, and I know how to make do with a lot. In any and all circumstances I have learned the secret of being content—whether well fed or hungry, whether in abundance or in need" (Philippians 4:11-12). The primary characteristic of maturity is to live consistently with one's beliefs (ch. $e^{i\pi}$). Paul says that he has trained and disciplined his own body and needs to make them be subject to his convictions, and he follows this statement with another famous verse, "I am able to do all things through him who strengthens me" (4:13).

The apostle Peter coaches us to grow spiritually: "Therefore rid yourselves of all malice, all deceit, hypocrisy, envy, and all

∫x/dx: Spiritual Maturity

slander. Like newborn infants, desire the pure milk of the word, so that you may grow up into your salvation if you have tasted that the Lord is good" (1 Peter 2:1–3). The psalmist calls out to the lost, "Taste and see that the Lord is good" (Psalm 34:8); and Peter says, if you have already tasted and seen that the Lord is good (that is if you have grasped the mystery of the gospel), then grow up into spiritual maturity. Furthermore, as we grow individually, Peter continues, "as you come to him, . . . you yourselves as living stones, a spiritual house, are being built to be a holy priesthood to offer spiritual sacrifices acceptable to God through Jesus Christ" (1 Peter 2:4–5). God is fulfilling his transcendent purpose in spiritually mature believers by building us into a priesthood (see ch. 7^3). What do priests do? They point other people to God; our purpose is to stand as a testimony to God's glorious wealth and immeasurable riches of grace that he bestowed upon us. When we point other people to the Savior as his priests, we are fulfilling the transcendent purpose to which God has called us; we cannot succeed in this mission without striving toward spiritual maturity through spiritual growth.

"But you are a chosen race, a royal priesthood, a holy nation, a people for his possession, so that you may proclaim the praises of the one who called you out of darkness and into his marvelous light" (1 Peter 2:9). God called us out of the darkness of believing that our lives were the product of meaningless chance, out of the Nothingness (ch. o), and into his glorious light, into his transcendent purpose (ch. *e*). This is why Jesus commands the woman "Go, and from now on do not sin anymore"; it is because she had been called up out of the meaningless slavery to her own passions and desires to fulfill God's glorious will in her life. Peter urges us to live in spiritual maturity, so that we can point others to Christ regardless of their points of view:

> Dear friends, I urge you . . . to abstain from sinful desires that wage war against the soul. Conduct yourselves honorably among the Gentiles, so that when they slander you as evildoers, they will observe your good works and will glorify God on the day he visits. (1 Peter 2:11–12)

How Can a Rational God Allow Irrational Numbers?

> For you were called to this, because Christ also suffered for you, leaving you an example, that you should follow in his steps. He did not commit sin, and no deceit was found in his mouth; when he was insulted, he did not insult in return; when he suffered, he did not threaten but entrusted himself to the one who judges justly. He himself bore our sins in his body on the tree; so that, having died to sins, we might live for righteousness. By his wounds you have been healed. For you were like sheep going astray, but you have now returned to the Shepherd and Overseer of your souls. (1 Peter 2:21–25)

Jesus sets a high bar for spiritual maturity. Ultimately, we will not reach his mark in this life. But, Paul tells the Philippians, "Not that I have already reached the goal or am already perfect, but I make every effort to take hold of it because I also have been taken hold of by Christ Jesus" (Philippians 3:12). This is the final reason to pursue spiritual maturity. Just as Peter told us to "grow up into your salvation if you have tasted that the Lord is good" (1 Peter 2:3), Paul tells us the reason he pursues godliness is because Jesus has taken hold of him and transformed him.

The author readily confesses that he lives in the conflict that Paul identifies in Romans 7:15–25. If the reader were to perform more than a cursory review of the author's life, he/she would easily identify a propensity for sinful behavior. There are still things in the author's life that are inconsistent with what the author believes. Maturity requires an individual to live consistently with the beliefs he/she holds, to align one's behaviors with one's beliefs (ch. $e^{i\pi}$). The author echoes Paul's words from Romans 7:24–25, "What a wretched man I am! Who will rescue me from this body of death? Thanks be to God through Jesus Christ our Lord! So then, with my mind I myself am serving the law of God, but with my flesh the law of sin." Furthermore, the author strives to live out Paul's policy in pursuing spiritual maturity: "Not that I have already reached the goal or am already perfect, but I make every effort to take hold of it because I also have been taken hold of by Christ Jesus" (Philippians 3:12). The author recognizes his battle with the sinful flesh but desires to live in the transformed life to which God has called him.

Chapter 3e

A^T: The Transformation

ACCORDING TO NEWTON'S FIRST law of motion (the law of inertia), an object in motion will stay in motion unless acted upon by an outside force, and an object at rest will stay at rest unless acted upon by an outside force.[1] This is the natural order of the non-quantum realm. Consider how this describes human complacency. Children usually have big dreams and plans about how they will change the world, but as they grow, this energy is diminished by the dragging frictional forces of education, hard work, overcoming peer pressure, their family circumstances, etc. For a good poetic reflection on this concept, listen to "Back When I Could Fly" by Trout Fishing in America.

In general, the average human settles for comfort over aspiring to reach childhood dreams. This is similar to the general principal of entropy in that these closed systems tend toward diminishing energy because of the inefficiencies of use.[2] Much of our energy is lost to the environment in the form of heat. This similitude is philosophical, not scientific, in nature, because we are not specifically considering thermal exchanges and processes.

---

1. Editors of *Encyclopaedia Britannica*, "Newton's Laws of Motion."
2. *Merriam-Webster Dictionary*, s.v. "Entropy."

How Can a Rational God Allow Irrational Numbers?

The author has often pondered what he has identified as psychological entropy. It seems that whenever we face challenges, they seldom come alone, and how we handle one challenge affects how we handle the next. Consider the chess game: If you were to block a bishop check with your knight, that knight becomes pinned to your king and is therefore not able to defend itself from an additional attack, nor is it available to ward off other attacks or effect an attack of its own. Once another check is delivered and another, this barrage of attacks leaves many of your pieces unprotected, and the game spirals out of control. In the same way, if you respond to a coworker in anger concerning a situation, your emotional well-being will be disrupted, and how you respond to your spouse later will likely be adversely affected. On the chessboard, a strong player can foresee the opponent's attacks and defend against them. In life, you cannot always be prepared for this psychological entropy. You wake up after a good night's rest rejuvenated, but an argument with your spouse or perhaps a breakfast accident soiling your pants initiates the day's progress to complete inner turmoil and chaos. Entropy decreases only when an outside force acts on the system; that is to say, an open system can receive energy from its environment, which causes its entropy level to decrease while the environment loses energy, causing its entropy level to increase.[3] Think of your life as an open system that is free to interact with the environment around you. You may be able to add energy by consuming food, getting a good night's rest, meditating on God's word, or having a calming conversation with someone who helps you gain perspective. In any of these events, there is a force exterior to your own self-system that is adding energy to reduce your psychological entropy.

So how can we bring balance to the force and to our lives? In the previous chapters we have talked about spiritual growth, the what, how, and why. We have talked about disciplining ourselves through experience and practice. Now let us consider the transformation that is occurring inside of us because of what God is doing.

In geometry, a transform is used to rotate, reflect, translate, skew, or dilate an image. These functions serve to alter the physical

---

3. Energy Education, "Entropy."

AT: The Transformation

configuration while preserving certain aspects of the original image. In other mathematics, computational transforms are used to simplify complex equations into manageable forms, so that operations can be performed on the simplified versions, and then the answer is transformed back into the original variables. One of the simplest examples of a computational transform is the substitution of a variable for a cumbersome factor to differentiate or integrate. Some of the more notable complex transforms are the Fourier and Laplace transforms, named for the mathematicians who introduced them. The process generally involves convolution and deconvolution.

God wants to transform us as individuals. In Acts 17, the Jews in Thessalonica were looking for the leaders of a new sect (the Christians) who were causing a disturbance in their nation. The Jews stated, "These men who have turned the world upside down have come here too" (v. 6). The Jews recognized that Christianity was bringing a radical change; they thought it was a convolution of nature, "turning the world upside down," as if the Christians had reflected the world over some natural axis through a geometric transform. However, if you look at this from God's perspective, humanity had already abandoned his plan; God is deconvolving our culture, removing the wickedness, the brokenness, and the deceit to transform it back to his perfect design.

The Bible tells us that Jesus came to be the Savior of the world (1 John 4:14), that Jesus is the light of the world (John 8:12), that Jesus came to testify to the truth (18:37), and that Jesus is the resurrection and the life (11:25). What do these imply about the world? The world is a lost, dark place, full of lies and death. God wants to deconvolve this world, transforming it back to his perfect creation. This work of transformation is performed one person at a time.

The psalmist tells us that it is the Lord that saves us from our darkness and bondage. "Then they cried out to the Lord in their trouble; he saved them from their distress. He brought them out of darkness and gloom and broke their chains apart" (Psalm 107:13-14). Paul tells the Corinthians in reference to the mystery of the gospel:

How Can a Rational God Allow Irrational Numbers?

> Therefore, since we have this ministry [transcendent purpose (ch. *e*)] because we were shown mercy [we have grasped the mystery of the gospel through salvation (ch. π)], we do not give up [we are growing spiritually (ch. 2*e*)]. Instead we have renounced secret and shameful things, not acting deceitfully or distorting the word of God, but commending ourselves before God to everyone's conscience by an open display of the truth [we are living in spiritual maturity, ch. 2π]. But if our gospel is veiled [a hidden mystery], it is veiled to those who are perishing. In their case, the god of this age has blinded the minds of the unbelievers to keep them from seeing the light of the gospel of the glory of Christ, who is the image of God. For we are not proclaiming ourselves but Jesus Christ as Lord, and ourselves as your servants for Jesus's sake. For God who said, "Let the light shine out of darkness," has shone in our hearts to give the light of the knowledge of God's glory in the face of Jesus Christ.
> (2 Corinthians 4:1–6)

This world is fallen and is influenced heavily by the one who deceives us, just as Adam and Eve were deceived by the serpent in the garden. We are still being deceived today. Paul tells the Ephesians, "Our struggle is not against flesh and blood, but against the rulers, against the authorities, against the cosmic powers of darkness, against evil, spiritual forces in the heavens" (Ephesians 6:12). How will God reconcile humans to himself? First, he must reveal himself, which he has done. He has shown a light out of the darkness. Now, he is working to transform us one by one.

Just after instructing the Philippians to put on the humility (ch. 42) that Christ has displayed (Philippians 2:1–11) and telling them to work out their own salvation (ch. φ), Paul says, "It is God who is working in you both to will and to work according to his good purpose. Do everything without grumbling and arguing, so that you may be blameless and pure, children of God who are faultless in a crooked and perverted generation, among whom you shine like stars in the world, by holding firm to the word of life" (2:13–16). God is transforming us from the inside out, but it is our job to partner with him in disciplining ourselves to develop and mature the

AT: The Transformation

divine nature, so that we can accomplish the transcendent purpose. In Romans, Paul tells us, "Therefore, brothers and sisters, in view of the mercies of God, I urge you to present your bodies a living sacrifice, holy and pleasing to God; this is your true worship. Do not be conformed to this age, but be transformed by the renewing of your mind, so that you may discern what is the good, pleasing, and perfect will of God" (Romans 12:1-2). Because you have grasped the mystery of the gospel through salvation, discipline yourselves to grow spiritually, to become spiritually mature in order to accomplish the transcendent purpose of God. Doing so will require you to transform your old way of thinking so you can grow spiritually.

To fully comprehend the transformation God is bringing, consider the philosopher Diogenes. Diogenes is called the father of cynicism because of his object lessons decrying the predictability of the selfishness of humanity with no potential cure. One of these object lessons entailed Diogenes walking through the center plaza of downtown with a lit lantern during the day. When the townspeople approached and asked what he was doing, Diogenes replied that he was searching for "an honest man."[4]

In the book of Isaiah, the prophet tells a similar story to that portrayed by Diogenes. The author encourages the reader to open his/her Bible to Isaiah 59 to read the entirety of this chapter, but for the sake of space and in the interest of the current point, a few verses are included here.

> [9]Therefore justice is far from us, and righteousness does not reach us.
> We hope for light, but there is darkness; for brightness, but we live in the night.
> [10]We grope along a wall like the blind; we grope like those without eyes.
> We stumble at noon as though it were twilight; we are like the dead among those who are healthy.
> [11]We all growl like bears and moan like doves.
> We hope for justice, but there is none; for salvation, but it is far from us.

---

4. Archon, "Philosophy of Diogenes," anecdote 24.

How Can a Rational God Allow Irrational Numbers?

¹²For our transgressions have multiplied before you, and our sins testify against us.
For our transgressions are with us, and we know our iniquities:
¹³transgression and deception against the Lord, turning away from following our God,
speaking oppression and revolt, conceiving and uttering lying words from the heart.
¹⁴Justice is turned back, and righteousness stands far off.
For truth has stumbled in the public square, and honesty cannot enter.
¹⁵Truth is missing, and whoever turns from evil is plundered.
The Lord saw that there was no justice, and he was offended.
¹⁶He saw that there was no man—he was amazed that there was no one interceding;
so his own arm brought salvation, and his own righteousness supported him.
¹⁷He put on righteousness as body armor, and a helmet of salvation on his head;
he put on garments of vengeance for clothing, and he wrapped himself in zeal as in a cloak.
¹⁸So he will repay according to their deeds: fury to his enemies, retribution to his foes, and he will repay the coasts and islands.

Notice that when God saw that we were searching for light (v. 9), groping in the darkness like blind men (v. 10), when God saw that honesty and truth were failing and falling in the public square (v. 14), when God saw that there was no one interceding (v. 16), he didn't become a cynic and sit back to criticize. God brought down salvation with his own arm, and his own righteousness supported him (vv. 16–17). The Christian's role in this world is not to sit back and condemn it. It is to partner with God in his reconciliation and deconvolution of it. The only way we can partner with God in his mission is to grow into spiritual maturity and then conform to his specific plans for us, becoming a variable to serve in his overall plan.

Chapter 3π

x: The Variable

VARIABLES ARE USED IN math equations to represent any number. Generally, these are defined as elements of some specific set. A proof may begin with, "Let x be an element of the integers (*let* $x \in Z$)." The set has defined properties. For instance, in the example, the set of integers consists of the whole numbers, positive, negative, and zero. In biology, we use a system for characterizing living organisms into kingdom, phylum, class, order, species, etc.[1] If we were playing a game of twenty questions, the first questions may center around the characteristics of the mystery.

When an archaeologist begins *digging* into ancient civilizations, he/she does not begin with a pickax and shovel. Long before a physical dig can initiate, study and research are performed to provide context. An anthropologist who enters an unfamiliar civilization for the first time will attempt to understand the cultures and historical context of the region before barging in. Why do these professions require so much study and understanding? It is perhaps similar to the reason the quantum realm does not obey Newton's laws of motion. Just as the electron behaves differently while being observed than while not being observed,[2] people's

1. Basic Biology, "Taxonomy."
2. Weizmann Institute of Science, "Quantum Theory Demonstrated."

How Can a Rational God Allow Irrational Numbers?

behavior changes when their patterns are disrupted by the very event of observation.[3] If you were to accompany the author to the grocery store, the odds of the author buying a package of Oreo cookies are significantly diminished. Why? Because the author knows he is being observed, and, psychologically, he equates observation with accountability. While the archaeologist may disturb or trample physical artifacts, the anthropologist may disturb or trample psychological or sociological artifacts.

While the professional archaeologist and anthropologist study their targets in order to avoid disturbing or changing them, Christ became one of us to identify with us, so that he could bring change to us. God is transforming us individually through Christ (ch. 3e). If we are to be effective variables or catalysts to fulfill God's transcendent purpose, we must study the audience where God has placed us, so that he can use us to bring this change, a spiritual awakening. The Word (Jesus) became flesh (human) and dwelt among us (John 1:14). Jesus is the variable God used to bring change into the world, and he is the example we are to follow (spiritual maturity).

If the reader were going to speak to the president of a large company, would you prepare beforehand? Would you research his/her company's mission, vision, and goals? Would you look at the latest press releases? Surely, the reader would at least find out something about the company's industry. How would you prepare for a meeting with a foreign diplomat? Would that be the same as your preparation for a casual conversation with your neighbor? Would you prepare for a formal dinner in the same way you would for a family barbecue? Our encounters in life require distinct levels and types of preparations. With an alternate series of questions, we could just as easily demonstrate that language, attire, mindset, and behavior are also dependent on the situations we face. We have learned to be variables to effectively navigate life's adventures.

The prophet Ezekiel was called to go proclaim God's message to the people of Israel: "Son of man, go to the house of Israel and speak my words to them. For you are not being sent to a people

---

3. Shuttleworth, "Hawthorne Effect."

X: The Variable

of unintelligible speech or a difficult language but to the house of Israel—not to the many peoples of unintelligible speech or a difficult language, whose words you cannot understand. No doubt, if I sent you to them, they would listen to you'" (Ezekiel 3:4-6). God proceeds to tell Ezekiel that while the people will not want to listen, God is equipping him with everything he needs. Why was Ezekiel sent? "If you warn the righteous person that he should not sin, and he does not sin, he will indeed live because he listened to your warning" (3:21). God transformed Ezekiel into a watchman and then a living object lesson. Ezekiel was not without objection. In chapter 4, when God told Ezekiel to lie on his left side for 390 days and then 40 days on his right side while bound with a cord, and that he was to take eight ounces of food each day, which would consist of barley cakes baked and dried over human excrement to signify the coming captivity, Ezekiel protested. God relented to let him use cow dung instead of human excrement. Later in chapter 5, Ezekiel shaved his head as a sign of the coming judgment.

God called Hosea to marry a whore, whom he later bought back from sexual servitude (Hosea 3). God called Samson and John the Baptist to be Nazarites. God doesn't call all believers to these extreme lifestyles, but he does call all believers to spiritual maturity and transformation to present the gospel to others. Paul tells us, "For everyone who calls on the name of the Lord will be saved. How, then, can they call on him they have not believed in? And how can they believe without hearing about him? And how can they hear without a preacher? And how can they preach unless they are sent? . . . So faith comes from what is heard, and what is heard comes through the message about Christ" (Romans 10:13-17). It is the transcendent purpose of the believer to grow into spiritual maturity and to glorify God by pointing others to Christ. Faith to salvation is evidenced by calling on the name of the Lord, but who will go and preach God's message? Our answer should be the same as Isaiah's upon beholding God's glory, "Then I heard the voice of the Lord asking: 'Who should I send? Who will go for us?' I said, 'Here I am. Send me'" (Isaiah 6:8).

How Can a Rational God Allow Irrational Numbers?

Paul, in his First Letter to the Corinthians, addresses the rights of the believers while discussing whether it is spiritually sound to eat meat offered to idols. While he concludes that it is not spiritually immoral in and of itself, he challenges us to not let our rights and knowledge as believers hinder the spiritual growth of others:

> Food will not bring us close to God. We are not worse off if we don't eat, and we are not better if we do eat. But be careful that this right of yours in no way becomes a stumbling block to the weak. For if someone sees you, the one who has knowledge, dining in an idol's temple, won't his weak conscience be encouraged to eat food offered to idols? So the weak person, the brother or sister for whom Christ died, is ruined by your knowledge. Now when you sin like this against brothers and sisters and wound their weak conscience, you are sinning against Christ. Therefore, if eating food causes my brother or sister to fall, I will never again eat meat, so that I won't cause my brother or sister to fall. (1 Corinthians 8:8–13)

Paul is advocating for our malleability to Christ's likeness over exalting our own rights and knowledge. Here is another challenging aspect of Christianity for the intellectual; spiritual maturity means submitting our own rights in humility (ch. 42) for the good of others.

Later in the same letter, Paul tells us that he is willing to become a variable according to the audience to whom God called him to preach:

> Although I am free from all and not anyone's slave, I have made myself a slave to everyone, in order to win more people. To the Jews, I became like a Jew, to win the Jews; to those under the law, like one under the law—though I myself am not under the law—to win those under the law. To those who are without law, like one without law—though I am not without God's law but under the law of Christ—to win those without the law. To the weak, I became weak, in order to win the weak. I have become all things to all people, so that I may by every possible

X: The Variable

means save some. Now I do all this because of the gospel, so that I may share in the blessings. (1 Corinthians 9:19-24)

Many years ago, the author was asked to teach a Spanish class to a group of adult church members. Learning a language is hard, especially as an adult, so it was encouraging to find that Christian adults were willing to spend their time learning even some basic communication skills in a foreign language. Before the class began the first day, the author overheard one person say something to the effect of "If they come to this country, they should learn to speak English." While on a fundamental level, the author agrees with this position for practicality and assimilation, it is important to note that until March 1, 2025, the United States had no official language.[4] Even so, the author addressed the class with something along these lines:

> I agree that people who wish to live and function in the United States should learn to speak English, but as you will come to appreciate, learning a new language requires an immense amount of time, energy, and dedication. Consider a new immigrant (whether legal or illegal) who does not yet speak English very well. What jobs are available to him/her? Cooking in restaurants, commercial cleaning, construction, and other similar jobs. These folks generally have families that they must clothe and feed as well. The work hours required to support these families, whether they are here or in their home countries, are generally long and arduous; they may work twelve to fourteen hours per day at one or more jobs, six to seven days per week. Imagine getting home to fix food for yourself or your family after working a twelve-hour day with a two-hour commute. When will you make time for a formal English class? Learning English on your own is a blind process requiring one to two hours per day of study and practice time. Who will practice with you since for economic convenience you likely live in a communal space with other non-English

---

4. USA, "Official Language."

How Can a Rational God Allow Irrational Numbers?

speaking people possibly including your family if they are here with you? While I will concede that people who live in the United States should learn English, I am not willing to say that until they learn English, they can go to hell. Thank God, Jesus doesn't require us to learn Hebrew to gain our salvation!

We must be willing to allow God to transform us into who he wants us to be. There is no better example of this transformation into a variable that God will use than that found in love. As intellectuals we would be wise to heed the following words from Paul in 1 Corinthians 13:

> If I speak human or angelic tongues but do not have love, I am a noisy gong or a clanging cymbal. If I have the gift of prophecy and understand all mysteries and all knowledge, and if I have all faith so that I can move mountains but do not have love, I am nothing. And if I give away all my possessions, and if I give over my body in order to boast but do not have love, I gain nothing.
>
> Love is patient, love is kind. Love does not envy, is not boastful, is not arrogant, is not rude, is not self-seeking, is not irritable, and does not keep a record of wrongs. Love finds no joy in unrighteousness but rejoices in the truth. It bears all things, believes all things, hopes all things, endures all things.
>
> Love never ends. But as for prophecies, they will come to an end; as for tongues, they will cease; as for knowledge, it will come to an end. For we know in part, and we prophesy in part, but when the perfect comes, the partial will come to an end. When I was a child, I spoke like a child, I thought like a child, I reasoned like a child. When I became a man, I put aside childish things. For now we see only a reflection as in a mirror, but then face to face. Now I know in part, but then I will know fully, as I am fully known. Now these three remain: faith, hope, and love—but the greatest of these is love.

Chapter 42

ε: Humility

YES, THIS CHAPTER NUMBER is a reference to the meaning of "life, the universe, and everything" as defined in Douglas Adams's *The Hitchhiker's Guide to the Galaxy*.[1] Epsilon in mathematics is generally associated with a tiny number. It is often used to denote a very small change in time or another independent variable and then to show how a dependent variable changes in relation to that small adjustment. In order to embrace faith and Christianity, the content of this particular chapter is vital. We must be willing to consider ourselves nothing in comparison to who God is. We must lay aside our own egos, which is no easy task for an intellectual.

In the game of chess, the king is the most important piece (and the weakest), and the object of the game is to protect him while capturing the opponent's king. One of the most highly esteemed tactics in the game of chess is the sacrifice. This tactic involves giving away a pawn or piece seemingly for free while effecting a counterattack to gain a material or positional advantage. Since the queen is the most powerful piece on the chessboard, when she is sacrificed, the moves are calculated (or desperate), and some decisive advantage is nearly always achieved.

---

1. Adams, *Hitchhiker's Guide to Galaxy*, 187–90.

How Can a Rational God Allow Irrational Numbers?

In the cosmic game of chess, the most important piece is also the most powerful piece, the King. We, the pawns, are weak but not expendable. As Jesus says, the Good Shepherd knows his sheep and will go after a single lost sheep (Matthew 18:12–14; John 10). God, in his divine sovereignty, sacrificed the King for the sake of the pawn. If a queen sacrifice for the sake of the king is extravagant in chess, how much more extravagant is God's King sacrifice for the sake of his pawns (Ephesians 1:7–9; 1 John 3:1)?

Once we have laid aside our pride at the cross and accepted salvation by God's grace (ch. $\sqrt{2}$), we must continually put our own pride behind us. Most of us are too busy attempting to appear as if we are the sort of chaps who "still know where our towels are"[2] to humbly receive the free gift of salvation, but in order to fulfill the transcendent purpose to which God has called us, we must be willing to submit ourselves to God's desires and plan. Perhaps we would be benefited by a visit to the Total Perspective Vortex, whereupon we would learn that we are not the center of the universe. If, however, we live in the alternate, virtual universes we (or our associates) carry in our own briefcases, we will be deceived, just as Zaphod Beeblebrox and so many others have been.[3]

The prophet Isaiah had an encounter with God in Isaiah 6, wherein he realized just who he was in relation to God. "I saw the Lord seated on a high and lofty throne, and the hem of his robe filled the temple" (v. 1). "Then I said, 'Woe is me for I am ruined'" (v. 5). When Isaiah found himself in the Total Perspective Vortex, he beheld the Master of the universe. That's when he realized, just as Zaphod was supposed to have observed, that his life and identity were meaningless in comparison. The psalmist says, "When I observe your heavens, the work of your fingers, the moon and the stars, which you set in place, what is a human being that you remember him, a son of man that you look after him?" (Psalm 8:3–4). Our lives are meaningless in comparison to who God is. Even in our intellectual circles, we speak somewhat unintelligibly about the vastness of the universe. Is it any wonder then that our lives are

2. Adams, *Hitchhiker's Guide to Galaxy*, 28–29.
3. Adams, *Restaurant at the End*, 61–88.

ε: Humility

insignificant in comparison to the Creator of the universe? Job also had an encounter with God Almighty when he questioned God. "Then the Lord answered Job from the whirlwind. He said: 'Who is this who obscures my counsel with ignorant words? Get ready to answer me like a man; when I question you, you will inform me'" (Job 38:1–3). As intellectuals, we are fond of pompously espousing facts and statistics to demonstrate the infinite depths of our own wisdom, but it is when God responds that we finally realize the volume of our tittle of knowledge (ch. *i*).

"Where were you when I established the earth? Tell me if you have understanding. Who fixed its dimensions? Certainly you know! Who stretched a measuring line across it? What supports its foundations? Or who laid the cornerstone while the morning stars sang together and all the sons of God shouted for joy?" (Job 38:4–7). While the reader may argue that this present society is vastly better informed than Job, the author is not ready to concede that we have come to rival God in his knowledge.

Paul tells us, "Brothers and sisters, consider your calling: Not many were wise from a human perspective, not many powerful, not many of noble birth. Instead, God has chosen what is foolish in the world to shame the wise, and God has chosen what is weak in the world to shame the strong. God has chosen what is insignificant and despised in the world—what is viewed as nothing—to bring to nothing what is viewed as something, so that no one may boast in his presence. It is from him that you are in Christ Jesus, who became wisdom from God for us—our righteousness, in order that, as it is written: Let the one who boasts, boast in the Lord" (1 Corinthians 1:26–31). Why does God choose the foolish, weak, insignificant, and worthless? God gives Paul this answer in a different context: "My grace is sufficient for you, for my power is perfected in weakness" (2 Corinthians 12:9). It is precisely our insufficiency that God is seeking. When I am able to provide for myself, I boast of my own accomplishments or possessions. When I am weak and destitute and God provides for me, that is when I will boast in God. God created us to be dependent on him. This is why Jesus tells us in Matthew 11:28–30, "Come to me, all of you

How Can a Rational God Allow Irrational Numbers?

who are weary and burdened, and I will give you rest. Take up my yoke and learn from me, because I am lowly and humble in heart, and you will find rest for your souls. For my yoke is easy and my burden is light." God does not expect you to reason or build your way into eternal life. He knows you cannot; and that is why he made the way for you (ch. $\sqrt{2}$). Receiving this requires a great deal of humility on the part of the intellectual.

If the reader grew up attending church in the United States of America, you may remember a song about a "wee little man" named Zacchaeus. He was so short that he could not see Jesus for the crowd of people surrounding him. What he lacked in stature, he compensated for in ego. He was a tax collector who cheated people out of money for a living. When he heard about Jesus, though, he set his ego aside. He climbed a tree to see Jesus because he recognized that Jesus was different. Jesus was the unique solution to his soul's burning questions. He was obviously not able to fill life's void with money. When he had an encounter with Jesus, he not only laid aside his ego, but he surrendered it altogether (Luke 19:1–9).

In the book of John, we meet a character named John the Baptist who, despite his awkward social skills and radical lifestyle, had built a following among the Jews. People were coming to him to be baptized because of his message of repentance. John the Baptist knew, however, that he was not the Messiah. Having understood his place, he didn't greedily latch on to his followers or to the ministry he had built. When one of his disciples told him, "Rabbi, the one you testified about, and who was with you across the Jordan, is baptizing—and everyone is going to him," John simply replied, "He must increase, but I must decrease" (John 3:26–30). That is the heart of spiritual maturity.

In Matthew 16, Jesus tells us, "If anyone wants to follow after me, let him deny himself, take up his cross, and follow me. For whoever wants to save his life will lose it, but whoever loses his life because of me will find it. For what will it benefit someone if he gains the whole world yet loses his life? Or what will anyone give in exchange for his life?" (vv. 24–26). In this statement, much like

ε: Humility

his teaching that the first will be last (20:16), Jesus is doling out wisdom to those who will listen. Our natural inclination is to fight for our own individual rights. That is the selfish, human nature as discussed in chapter $\sqrt{2}$; but is this good, or does it have intrinsic value? Let us divorce ourselves momentarily from the Christian worldview presented herein and consider the *natural* world with all of its contradictions and conflicts once more.

Consider our present society. Some groups advocate for an individual's rights to govern their own body to the peril of society, while alternate groups advocate for an individual's rights to govern their own possessions to the peril of society. The individual says, "I want to do what makes me happy regardless of the consequences to society." As an example, one man determines to engage in sexual intercourse with one hundred women without concern for the potential progeny that may come from these engagements. Let us assume that the women in these engagements also share his free spirit, and the encounters result in no emotional baggage or damage. If some of the women choose to bear children from these encounters, the children unknowingly share DNA. When these children mature sexually and, in the cavalier free spirit of their parents, pursue various sexual partners, there is an exponentially increased chance that half siblings, unbeknownst to them, will engage with each other. The biological offspring of these multiple relationships will compound over generations. Is this *good* for society?

This is precisely why Christ brings transformation and calls us to live differently. Christianity instructs us to stop focusing on our selfish interests. Christ himself chose to willingly sacrifice his own present comfort for the long-term good of humanity, and Christianity teaches us to follow Christ's example. Christianity also involves discipline in the form of submission to Christ's way of thinking. This is why Paul tells us that spiritually mature believers "take every thought captive to obey Christ" (2 Corinthians 10:5). It is not sufficient to change behaviors; Christianity is a philosophical change in the way we see the world (ch. $e^{i\pi}$) and a spiritual change in the way we seek to understand the world (ch. 2π).

How Can a Rational God Allow Irrational Numbers?

In Philippians, Paul instructs us as Christians to adopt Christ's attitude of a humble servant: "Adopt the same attitude as that of Christ Jesus, who existing in the form of God, did not consider equality with God as something to be exploited. Instead he emptied himself by assuming the form of a servant, taking on the likeness of humanity. And when he had come as a man, he humbled himself by becoming obedient to the point of death—even to death on a cross" (Philippians 2:5–8). In human governments, leaders are set on pedestals and rarely mingle with commoners. In big corporations, it is the same. The author considers it a privilege to work in an organization where the president of the company is not only able but also willing to perform basic, everyday tasks such as getting coffee for others, making copies, and tidying up the conference room after a meeting. The reader could respond that the president gets paid too much to spend time on these menial tasks. However, one must remember that the most important name on a business card is not that of the employee, which is typically found in the center, but rather that of the company, which is typically found at the top. The president's behavior reflects on the company as a whole. Elitist thinking on the part of the president begets elitist thinking on the part of employees. Whenever we begin to believe that certain tasks are beneath us or our position, we have lost the humility required to value our clients, coworkers, cleaning crew, and other people in general over ourselves.

Just as the president's actions reflect on a company, any given employee's actions represent the company. As John Donne said, in contradiction to the much later Simon and Garfunkel song, "No man is an Island."[4] The work of mass murderers is a reflection on humanity as a whole just as are the actions of the altruistic philanthropist. As Shakespeare said, "All the world's a stage,"[5] but we judge the performance by the quality of the actors. On the world stage, America is judged by the actions of her actors, those who travel abroad, those who are in the news, and those who are on social media. When the author lived in Lima, Peru, people in certain

---

4. Donne, *Devotions*, meditation 17; Simon, "I Am a Rock."
5. Shakespeare, *As You Like It*, 2.7.31.

ε: Humility

areas of the city constantly offered him drugs, tattoos, and sex. The image of Americans that the Peruvians had was drastically different than the values the author possessed (and still possesses). Call it prejudice or stereotyping if you like; while we may agree that it is unfair or, at the very least, unrepresentative of the individual member, it still happens. We, as humans, make similar judgment calls every day. We do not let blind people drive vehicles; is this unfair profiling? While we may wish to divorce ourselves from the actions of our president or other famous Americans, the reality is that even if we do not agree or approve of their actions, they do represent America and her values. Unfortunately, the most memorable and therefore readily established formative control points are found at the extremities and are almost exclusively negative in nature. Christianity is no different. While we should point to Christ as the principal example of the Christian faith, the behavior of any given Christian is also viewed as and used by the general public to form an image of what Christianity is. The author has often lamented, "If it weren't for Christians, there would be more of us." The author says this in criticism of himself.

Regardless of whether humanity as a whole subscribes to Christianity, humanity would be greatly benefited by imitating the selflessness of Christ as demonstrated by his humility. When the individual esteems the corporate good more highly than his/her own rights, society benefits. But how can we promote humility? Without esteeming Christ as the ultimate example and subscribing to the Christian worldview, humans are naturally prone to exalt themselves, as we have discussed in chapter $\sqrt{2}$, and left without boundaries in defining their identities, as shown in chapter $e^{i\pi}$. Individuals could learn and practice humility as a discipline without Christ. It is possible that an individual could accept option 1b presented in chapter 0 to "serve the nonunique 'greater good' of humanity, the universe, the earth, or some other ultimately meaningless cause exterior to oneself," but there would be no foundation and no ultimate driving force to provide motivation, no response to the prevalent question: *Why?*

How Can a Rational God Allow Irrational Numbers?

In Christianity, we find Christ gives us the purpose for living in humility, as discussed in chapters e and π; the means to gain humility, as discussed in chapters $2e$ and 2π; and the particulars of how to use humility to achieve our purpose, as discussed in chapters $3e$ and 3π. If we esteem humility a valuable trait, the Christian worldview is a very logical conduit for proliferating it in society.

Consider this question. What could you possibly submit or surrender to compare with what Jesus has sacrificed? Or, what do you possess that is so precious to you that it is more valuable than your relationships with others? Jesus humbled himself from Creator to creation; he took on our limitations, our life cycles, our needs, and our weaknesses. Regardless of your position, no matter the degree of your submission, there is no comparison to the humility Christ displayed. For humans, the greatest single limiting factor to building and maintaining relationships with others is our pride.

When humans engage in sporting competitions, we strive to be the best, desiring to obliterate opponents to show how great we are. When we lose, our egos are crushed. And when our opponents offer sincere consolation, we become too proud to accept the proffered handshake. Why is that? Which is worth more: the pride in the skill or the skill itself?

As intellectuals we tend to believe that we have aspired on our own to become the men and women we are today; we have gained knowledge and wisdom through our own efforts; we have pulled ourselves up by our own bootstraps. That is the source of our pride, but it is unfounded. Society has contributed to our growth. Our parents have contributed to our growth. Our friends, teachers, doctors, news anchors, garbage collectors, realtors, grocery attendants, etc. have contributed to our growth. Without others, we are nothing. Forget for a moment that you would not exist without a mother and father (or perhaps a lab scientist, test tube, and a donor); you were helpless in your formative years. When did you learn to communicate, feed yourself, dress and protect yourself, walk, etc.? Without the infrastructure and time of others, you would not be who you are today (ch. 7^3).

ε: Humility

Paul, having been transformed by his encounter with Christ, told the Philippians, "Everything that was a gain to me, I have considered to be a loss because of Christ. More than that, I also consider everything to be a loss in view of the surpassing value of knowing Christ Jesus my Lord. Because of him I have suffered the loss of all things and considered them as dung, so that I may gain Christ and be found in him, not having a righteousness of my own from the law, but one that is through faith in Christ—the righteousness from God based on faith" (Philippians 3:7–9). Until we are willing to surrender our own accomplishments, egos, and self-interests, we cannot know the power of God. True Christianity is to have a heart like Christ; it is to humbly submit oneself to the God of the universe.

Jesus addressed the subject directly: "The disciples came to Jesus and asked, 'So who is the greatest in the kingdom of heaven?' He [Jesus] called a child and had him stand among them. 'Truly I tell you,' he said, 'unless you turn and become like children, you will never enter the kingdom of heaven. Therefore whoever humbles himself like this child—this one is the greatest in the kingdom of heaven'" (Matthew 18:1–4). Later when two of his disciples, James and John, asked to rule with him at his right and left sides, Jesus responded, "You know that those who are regarded as rulers of the Gentiles lord it over them, and those in high positions act as tyrants over them. But it is not so among you. On the contrary, whoever wants to become great among you will be your servant, and whoever wants to be first among you will be a slave to all. For even the Son of Man [Jesus himself] did not come to be served, but to serve, and to give his life as a ransom for many" (Mark 10:35–45). To exalt Christ is to humbly serve those around you; it is not human nature to be a servant, and that is precisely why God is glorified—because this is a supernatural attitude.

In Aldous Huxley's *Brave New World*, in a future time, a class system is established to create utopia (spoiler alert: really a dystopia, as revealed later in the book). Humans are bred in laboratories to fill certain roles in society. The Alphas, Betas, Gammas, and Epsilons each have their own assignments, and they are conditioned to not mingle with people of other classes. The role of the Epsilon is the servant to all the other classes. They are genetically designed to serve the other classes, and their function in society is ingrained in them from when they are babies. They never question why or aspire to rise to a new class. The Epsilon in Brave New World is small and invisible, performing the role of a servant without regard to his/her own life, purpose, or value.[6] This is not what God calls us to. God demonstrated our infinite value through his willingness to sacrifice his own life for us. We are not called to self-denial for the sake of fitting into society; we are called to selfless service to bring transformation to society. We are not called to devalue our own lives; we are called to value the lives of others because of God's infinite love toward us and toward them. We are not called to mingle only with those of our class or mindset; we are called to be fishers of men engaging all walks of life. Being an Epsilon in the kingdom of God means that we set aside our own agendas to fulfill that of Christ. "I must decrease, and he must increase."

---

6. Huxley, *Brave New World*, 222–23.

Chapter 7³

[Set Theory]: The Church

THIS PARTICULAR CHAPTER NUMBER is a nod to Andrew Beal's conjecture[1] that followed Fermat's last theorem. Fermat's last theorem is a famous math puzzle that remained unresolved until the last decade of the twentieth century.[2] At that same time, the author was a mathematics student at Northwestern State University in Natchitoches, Louisiana, and learned of Beal's conjecture during work on his final project. Beal conjectures that there are no solutions to the equation $A^x+B^y=C^z$, where A, B, and C are elements of the set of natural numbers and coprime with one another, and where x, y, and z are elements of the integers greater than two. To date, the author has not seen a solution to Beal's challenge or a proof that there is no solution. The chapter number is part of one equation ($10^2+3^5=7^3$) that is a near solution (one of the exponents is 2 in this example) that the author has identified.

We have Georg Cantor to thank for set theory. He was a brilliant mathematician who was severely criticized by the educated mathematical community of his day. While today much of his work may be considered canonical to certain branches of high-level mathematics, his contemporaries drove him to depression

---

1. See https://www.bealconjecture.com/.
2. Editors of *Encyclopaedia Britannica*, "Fermat's Last Theorem."

and near mental anxiety.[3] This is yet another example of how the educated (scientific) community practiced unfounded faith in what was already known to reject truth that was in the process of being discovered (ch. i).

Set theory is concerned with defining the nature of a group of numbers (set) and their general behavior, operations, and properties without regard to the specific elements of the set itself. For instance, the natural numbers is a set that is closed on addition and multiplication, but it is not closed on subtraction or division. Since when two natural numbers are added or multiplied together, the result is always also an element of the set; these operations are defined for this set. Since there exist at least two elements, x and y, of the natural numbers such that x minus y is not an element of the set, subtraction is not defined for this set. Similarly, since there exist at least two elements, x and y, of the natural numbers such that x divided by y is not an element of the set, division is not defined for this set.

How does this apply to Christianity? There are different subsets within the greater set of Christianity. Christianity, defined as the church, is the body of Christ across the globe and throughout history, in which all believers are united. For the purposes of this discussion, let us define this set as the global church. Within this set, there are various denominations and even congregations within these denominations, that are defined as churches. The global church should be viewed as the complete set containing the subsets of various denominations, which contain the smaller sets of congregations. There are differences within the denominations and perhaps even in the churches within the denominations. For the purposes of this discussion, let us define these elements of the global church as the local church. What are the functions of the global church and local churches?

First, let us consider what Jesus meant when he spoke to Peter about building his *church*. In Matthew 16:13–20, we find Jesus talking with his disciples. He asked, "Who do people say that the Son of Man [referring to himself] is?" The disciples responded with

3. Editors of *Encyclopaedia Britannica*, "Georg Cantor."

[Set Theory]: The Church

various theories. Then Jesus asked, "But you, who do you say that I am?" Peter responded, "You are the Messiah, the Son of the living God." When Jesus heard this, he said, "Blessed are you, Simon son of Jonah, because flesh and blood did not reveal this to you, but my Father in heaven. And I also say to you that you are Peter, and on this rock I will build my church, and the gates of Hades will not overpower it. I will give you the keys of the kingdom of heaven, and whatever you bind on earth will have been bound in heaven, and whatever you loose on earth will have been loosed in heaven."

This passage causes disagreement between various Christian groups, but it has birthed many funny jokes about meeting Peter at the "pearly gates," attempting to enter heaven. The author does not wish to dissect Jesus's words at great length; however, a limited discussion is warranted. Is Jesus saying that Peter, which means "rock," is the foundation of the church? Is Peter then the patriarch of the Christian church? Since Abraham is the patriarch to whom the promise of the Messiah is made and from whom the Messiah is to come, Peter is not the patriarch of Christianity. Peter also cannot be the foundation of Christianity because, in his own sermon in Acts 4 (see v. 11), Peter proclaims that Jesus has become the cornerstone, referencing the Old Testament prophecies found in Psalm 118:22 and Isaiah 28:16. Peter repeats this reference in 1 Peter 2:6–7. The latter of the two Old Testament references, Isaiah 28:16, says, "Therefore the Lord God said: 'Look, I have laid a stone in Zion, a tested stone, a precious cornerstone, a sure foundation; the one who believes will be unshakable." In Ephesians 2:20, as we will discuss briefly later, Paul tells us that we are being built into a temple "on the foundation of the apostles and prophets with Christ Jesus himself as the cornerstone." So, Peter is not the foundation itself nor the patriarch of the church. He is also not the principal, because Paul tells us in Romans 8:29 that Jesus is the firstborn among many brothers and sisters (as we will see, the local church is a family).

Furthermore, Paul tells us in 1 Corinthians 3 that assigning such prominence to individuals other than Christ is immature; remember that we want to grow into spiritual maturity by thinking, acting, and living like Christ. "For my part, brothers and sisters, I was not able to speak to you as spiritual people but as people of the flesh, as babies in Christ. I gave you milk to drink, not solid food, since you were not yet ready for it. In fact, you are still not ready, because you are still worldly. For since there is envy and strife among you, are you not worldly and behaving like mere humans? For whenever someone says, 'I belong to Paul,' and another, 'I belong to Apollos,' are you not acting like mere humans?" (vv. 1–4). Paul specifically says that this behavior is worldly because it creates envy and strife. He clarifies that the only foundation is Jesus Christ. "What then is Apollos? What is Paul? They are servants through whom you believed, and each has the role the Lord has given. I planted, Apollos watered, but God gave the growth" (vv. 5–6).

What is Jesus saying then? Remember that faith is like proof by induction (ch. i^i). The faith of those who have gone before us substantiates our faith. The apostles and prophets demonstrated their faith, and they form part of the foundation of our faith. Their faith was rooted and grounded in Jesus Christ, the cornerstone. Those who lived before Christ placed their faith in the promise that was to come; we place our faith in the fulfilled promise in Christ. Our faith, as we grow and mature in Christ, will become part of the foundation for our fellow believers and our children. Peter acknowledges, in faith, that Jesus is the Messiah; in spite of the doubters, scoffers, and popular opinion, Peter boldly proclaims, "You are the Messiah." Jesus responds by affirming Peter's faith, and it is upon this rock of faith that Christ will build his church. Why is that important? Because the church is based on the faith of individuals. In order to be a part of the church, one must first profess faith in Christ Jesus. The church comprises the individuals called to faith in Christ. The gates of Hades will never overpower the one who professes faith in Christ; the keys to the kingdom of heaven are given to all believers, not just to Peter, because the one believes in the sure foundation of Christ, the cornerstone, will be unshakable.

When Christ said he was going to build his church on this rock, he was, in fact, referring to both the global church and the local

[Set Theory]: The Church

church. Now let us proceed with a description of the global church, which we have defined as the worldwide, timeless body of Christ. A body is generally formed of many individual members, which perform unique functions (consider your hands, feet, heart, liver, eyes, ears, blood, skin, etc.). Each Christian becomes a member of this set, the global church, when he/she accepts Christ's sacrifice for his/her sins (ch. $\sqrt{2}$). The Christian then gains the properties belonging to this global set of Christians, including the transcendent purpose (ch. e), as well as access to the mystery (ch. π). New Christians are reborn into a new dimension of spiritual growth (ch. $2e$), into spiritual maturity (ch. 2π). And every new Christian is being transformed (ch. $3e$) into the variable (ch. 3π) God is using to reach others. Every Christian has these characteristics, though we are at different ages and stages in our spiritual growth.

What is the importance of the global church? The global church provides a structure for understanding God's revelation, God's character, and God's transcendent purpose for humanity. It is a large community of people at various stages of life functioning as a network to help grow and mature newer, younger believers. As an individual lives and functions in secular society, community provides opportunities to work, serve, eat, and participate in events. In a community, individuals fulfill the fundamental role of a mature person and identify with others in either similarities or differences. The global church serves an analogous purpose in Christian society, offering the believer opportunities to meet, work, serve, fellowship, etc., ultimately fulfilling the transcendent purpose to which God has called him/her. Community is comprised of family units; it is not composed merely of individuals.

What is the role of the local church? If the global church is a community, then the local church is a family. Your family knows you more intimately than does your community. Your family may also have more specific properties, such as identified mealtimes, a curfew, a bedtime, etc. While on occasion an individual may eat as part of the community, the majority of his/her alimentation is with the family unit. The family unit is responsible for raising children to properly function in the community. Remember, we loosely defined maturity as becoming a contributing member of society, and spiritual maturity as how Christlike someone is. The family is responsible for helping the individual become mature (chs. $2e$ and 2π). While these properties are inherited from belonging to the global church, they are honed and refined within the local church. For instance, while being born in the United States may grant the individual American citizen certain rights, the individual American citizen learns what these rights are and how they are expressed within subsets of society and in the family.

The most critical need of a newborn baby is a family who gives love, protection, and provision. The newborn is not consciously aware of community and likely does not comprehend or value family. In the same way, a newborn Christian may or may not be consciously aware of or appreciate the global church or the local church. Does this lack of awareness or appreciation negate the newborn baby's need for family and community or the newborn Christian's need for the local church and the global church? As the believer participates in the family (local church) and the community (global church), he/she is being transformed by God's presence into his character to become his servant and representative based on the audience to which God has directed him/her (chs. $3e$ and 3π).

In Ephesians, we learn that salvation comes by grace through faith (2:8–9), but later in the same chapter, Paul tells us that while we were strangers to God, he made us members of his household (global church/community and local church/family). That household is built on the foundation of the apostles and prophets, with Christ Jesus as the cornerstone (2:19–22). Faith requires a basis, as discussed previously (ch. i'). In the book of Acts, the members of the early local church met together daily to teach and learn and to encourage one another. They had all things in common. This was not communism as we have in political systems today, wherein the value of the individual is derived from the community as he/she serves the needs of the community. In Christianity, the individual's value stems from God, not from the organization or institution of the church. The early church had all things in common as community, a *commonism*, where each member is valued by every other member because the Creator values the individual member.

If the ultimate goal is for us to become contributing members of the community or spiritually mature members of the body of Christ, why then do we need the local church, the family? James 1:27 tells us, "Pure and undefiled religion before God the Father is this: to look after orphans and widows in their distress and to keep oneself unstained by the world."

The latter portion of this verse makes religion a personal responsibility. The author must keep himself unstained by the world. While we are growing and learning, we depend heavily on the family for our basic needs, such as the identification of healthy foods. In Christianity, that dependency is just as real. As young believers, we need other believers to help feed and disciple us. At some point, however, we should be more responsible in studying God's word on our own; but even then, we should meet together to discuss meaning and life application, just as the reader may meet for business meetings, trainings, or conferences in his/her profession. Once a child has become an adult, he/she must accept responsibility for his/her own life and choices; we cannot live with our parents forever. The maturing Christian must also be responsible for his/her own spiritual development. God makes himself known individually to each person; in Jeremiah 31:33, Hebrews 8:10, and Hebrews 10:16, God declares that he will write his laws on our hearts so that we will be his people. This level of intimacy from God demands a personal dedication from individual Christians, not just a corporate obeisance. The local church, composed of other Christians on their personal paths toward spiritual maturity, fosters the proper support and necessary guidance for the new Christian to grow into his/her personal relationship with Christ and to take responsibility for his/her own spiritual growth on the path toward spiritual maturity.

What about the first part of James's definition of religion: caring for the orphans and widows? It is in the local church that we find the opportunity to meet the needs of those around us. Just as some families in secular society require assistance from others, in the Christian community, there are individuals with both physical and spiritual needs. Even an infinitely wealthy philanthropist could not appropriately administer aid to all the financial needs in the United States; he/she would need local administrative bodies to evaluate and validate needs and then manage the distribution of funds. If a single Christian is isolated from the family, his/her ability to tend to the needs of others is severely hampered, and the administration of doing so becomes burdensome and inefficient. Besides this, as members of a body, we have different gifts and abilities (Romans 12:3–8; 1 Corinthians 12). So, we benefit from participating with the family because our weaknesses are complemented by another's strengths, and our strengths complement the weaknesses of others. We cannot do this alone.

Participation in the family (local church) spurs us on to Christlikeness (spiritual maturity). Hebrews 10:23–25 tells us not to neglect gathering ourselves together, so we can "consider one another in order to provoke love and good works." We meet together in the church to encourage each other in Christ. We need fellowship within our local body to help us mature by improving consistency.

While Christianity is personal and private, it is also practiced inside one's family (local church) and community (global church). Since faith is based on the accounts of others, we need brothers and sisters in Christ to share their experiences with us, just as professionals require colleagues. Proverbs 27:17 tells us that iron sharpens iron, and one person sharpens another. Ecclesiastes 4:12 tells us that one person can be overcome, but two people are stronger, and three are stronger still. We need others to help us grow spiritually and to help us achieve our transcendent purpose. Beyond that, others need us to invest in their journey of faith as well. Whom are you helping grow spiritually? Whom are you encouraging in the faith to achieve their transcendent purpose? While Christianity begins as an individual finding redemption in the love and sacrifice of Christ, it extends when that same individual then begins to love and sacrifice to help reach others. That is the function of the global church and the local church.

Chapter ∞

QED: Live What You Believe

WE HAVE COME TO the end of the "proof," quod erat demonstrandum (QED). This book is no scholarly paper presenting a new Euclidean proof of some heretofore unsolved mathematical postulate; it is a proof for the rationality of faith in the God of the Bible. This book was written to demonstrate the plausibility and reasonableness of the Christian worldview. In Acts 26, King Agrippa responds to Paul's arguments in favor of Christianity, asking, "Are you going to persuade me to become a Christian so easily?" Paul counters, "I wish before God that whether easily or with difficulty, not only you but all who listen to me today might become as I am" (vv. 28–29).

Jesus uses a phrase very similar to QED on the cross, when he says, "It is finished!" (John 19:30). What had he completed, or what had he proved? How should we respond?

As we mentioned in chapter *e*, God's transcendent purpose for us includes demonstrating "the immeasurable riches of his kindness to us in Christ Jesus," since "we are saved by grace through faith" (Ephesians 2:6–8). When Jesus proclaimed, "It is finished!" on the cross, he had completed the proof of God's eternal kindness and love toward us. We could summarize Christ's life and sacrifice for us as a brute-force method of proof. Instead of using deduction or induction, Jesus tested and was tested in every possible way

How Can a Rational God Allow Irrational Numbers?

to demonstrate that love. God left no room for doubt. Remember how irrational it is that the just would die for the unjust (Romans 5:6–11 [ch. $\sqrt{2}$]). What is it about our identities (ch. $e^{i\pi}$) that makes us worthy of the Creator's sacrifice? Since it was finished on the cross, why is it important that Jesus rose from the dead? This is another mystery of the gospel (ch. π).

Theologically, Christ's resurrection is important, because when Jesus rose from the dead, he conquered death, thus overcoming sin, which is the cause of death. Christ's resurrection is critical to Christianity because it is the promise of our hope and future in eternity. Rationally, if Christ did not rise from the dead, then we are fools to believe in the transcendent purpose to which God has called us. In fact, Paul makes this argument in 1 Corinthians and concludes, "If we have put our hope in Christ for this life only, we should be pitied more than anyone" (15:19). Jesus's death and resurrection were necessary and sufficient for salvation and redemption, but they also solidify the very desire at the heart of humans, a sense of belonging or importance, a transcendent purpose (ch. e).

In light of these events, what is our response? Should we continue living just as we did before? Paul tells us in Romans that such a response is absurd. He goes further to say, "If we have been united with him in the likeness of his death, we will certainly also be in the likeness of his resurrection" (Romans 6:5). We are called to grow spiritually (ch. $2e$), to become spiritually mature (ch. 2π). We are not brought to life through faith to continue the same meaningless existence we had before (ch. 0). In fact, James tells us that "faith without works is dead" (James 2:26). If God has brought us to life in Christ and thereby given us a transcendent purpose, why would we continue to live as if we are dead? While Christ's actions on the cross completed the work to prove God's immense love for us, his resurrection grants us the transcendent purpose; we must proceed to fulfill that transcendent purpose.

In John 8, Jesus instructed the woman caught in adultery to begin a new life: "Go, and from now on do not sin anymore" (v. 11); he gave her meaning. He doesn't save us to leave us in our

QED: Live What You Believe

meaningless happenstance of an existence. Jesus didn't call this woman out; he called her up. Just as God called her up to a new purpose, God is working to transform each individual believer into Christlikeness such that our character, understanding, worldview, and behavior reflect that of Christ (ch. 3e). But God is also in the process of transforming or redeeming the world by using the individual believer who is functioning within the cultures of the world (ch. 3π). In order to effectively participate with God in this transcendent purpose, I must be willing to lay aside my own selfishness, pride, and ego; submit my own passions, desires, and agendas; and be obedient to his direction (ch. 42). Should I do this all alone? Does God bring us to life and then expect us to immediately learn to care for ourselves? No, he also provided community and family for us in the form of the global church and the local church, respectively (ch. 7^3).

The book of Ecclesiastes is an account of King Solomon's attempts to seek purpose or meaning in this life. Solomon is commonly believed to have been the wisest and wealthiest person to have ever lived. He pursued labors, riches, power, sex, and, yes, even knowledge. He applied the scientific method by experimenting, measuring, recording, and repeating the experiments. Solomon found that all of his experiments failed to satisfy his search for significance (ch. 0). "Absolute futility. Everything is futile" (Ecclesiastes 1:2). In the end, King Solomon concluded, "There is no end to the making of many books, and much study wearies the body. When all has been heard, the conclusion of the matter is this: fear God and keep his commandments, because this is for all humanity. For God will bring every act to judgment, including every hidden thing, whether good or evil" (12:12–14).

Just like King Solomon, each of us is seeking our personal identity (ch. $e^{i\pi}$), the mystery of our own existence, the purpose of our own lives, a significance beyond this life. Where would we be as a society today if we did not listen to the experiences and build on the experiments of those who have gone before us (ch. i^i)? The selfish egotist believes that his opinion alone is worth holding and no one can teach him/her anything. A wise person, however,

How Can a Rational God Allow Irrational Numbers?

listens and carefully considers the experiences of others and incorporates third-party, additional information into his/her own life. Only a fool ignores the lessons of the past.

As stated in ch. 42, we must be willing to humble ourselves, realizing that the universe does not revolve around us, and embrace the transcendent purpose that God has given us. In Philippians 3, Paul sets an example for us when he says, "Not that I have already reached the goal or am already perfect, but I make every effort to take hold of it because I also have been taken hold of by Christ Jesus" (v. 12). Paul recognized his own insufficiency and embraced God's call. We will return to this in a moment, but consider Paul's warning, beginning in verse 18: "For I have often told you, and now say again with tears, that many live as enemies of the cross of Christ. Their end is destruction; their god is their stomach; their glory is in their shame; and they are focused on earthly things" (vv. 18-19). Paul's words could easily describe the intellectual of our day and time. Our god is our own stomachs, egos, bank accounts, reputations, etc. We live as enemies of the cross because we reject Christianity as foolishness. Our glory is the things we boast of that are meaningless, like our passing knowledge. Once we die, our specific knowledge dies with us; regardless of whether we write these down or teach them to others; one's specific insights are lost forever to this existence. "For the word of the cross is foolishness to those who are perishing, but it is the power of God to us who are being saved. For it is written, I will destroy the wisdom of the wise, and I will set aside the intelligence of the intelligent" (1 Corinthians 1:18).

Paul coaches us as his family (ch. 7^3): "Brothers and sisters, I do not consider myself to have taken hold of it. But one thing I do: Forgetting what is behind and reaching forward to what is ahead, I pursue as my goal the prize promised by God's heavenly call in Christ Jesus" (Philippians 3:13-14). Paul's goal is to fulfill the transcendent purpose God has given him, and he admonishes us, "Therefore, let all of us who are mature think this way" (3:15a). First, he appeals to our intellect, reasoning that what is behind us cannot be changed, and he asks us to be mature in our thoughts

QED: Live What You Believe

(ch. φ). Then he encourages us to live like him: "Join in imitating me, brothers and sisters, and pay careful attention to those who live according to the example you have in us" (3:17).

As noted in chapter 1, the existence of a supernatural Deity, God, is arguably the simplest solution to the multivariate problems involving why humans seek purpose, why we have a sense of justice and shame, and why we have any internal moral code. As we concluded in chapter 0, if there is no God, then there is also no universal purpose to existence. In the absence of a universal purpose to our existence, studying mathematics, physics, chemistry, biology, astronomy, astrophysics, astrology, history, philosophy, engineering, morality, psychology, sociology, language, culture, or anything else is ultimately meaningless to the individual and even to the species. The unconscious forces of nature will defeat her creation even though it has become conscious. Mother Nature will win in the end. If there is no God, then there is also no absolute morality. There is no moral reason to protect endangered species or weaker members of our own species. In fact, we are fighting Mother Nature when we do so. Murder and slavery should not be considered bad; they should be embraced for the natural evolutionary processes they are.

Anyone who believes there is no God and continues to seek to gain more knowledge or continues to promote protection of the environment or others for any end other than self-promotion is a blundering fool. Of Christians, Paul says, "If we have put our hope in Christ for this life only, we should be pitied more than anyone" (1 Corinthians 15:19). For the atheist, this passage may read: "If we put our hope in humanity and the universe for a time beyond this life, then we should be pitied more than anyone." It is the author's opinion that maturity requires one to live in a manner consistent with one's own beliefs. If one's behaviors contradict one's beliefs, he/she should examine his/her behaviors and conscience. If these latter prove desirable, then he/she must necessarily question his/her beliefs and the foundations of those beliefs (ch. $e^{i\pi}$). To do otherwise is actually to practice blind faith, because there is no underlying purpose to warrant the behaviors (ch. φ). As an intellectual,

How Can a Rational God Allow Irrational Numbers?

I challenge you to substantiate your beliefs of a credible worldview and to live it with consistency. James tells us, "But someone will say, 'You have faith, and I have works.' Show me your faith without works, and I will show you my faith by my works" (James 2:18). What do you believe?

QED

Bibliography

Adams, Douglas. *The Hitchhiker's Guide to the Galaxy*. Hanomag, 2001.
———. *The Restaurant at the End of the Universe*. Del Rey, 2005.
Ainsworth, Thomas. "Form vs. Matter." *Stanford Encyclopedia of Philosophy*, Feb. 8, 2016; last updated June 27, 2024. Edited by Edward N. Zalta and Uri Nodelman. https://plato.stanford.edu/entries/form-matter/.
Archon, Sofo. "The Philosophy of Diogenes." Sofo Archon, n.d. https://sofoarchon.com/the-philosophy-of-diogenes/.
Aristeidis9. "History of Blindfold Chess." Chess, last updated Feb. 21, 2009. https://www.chess.com/article/view/history-of-blindfold-chess.
Bailey, David H. "Do Probability Arguments Refute Evolution?" Science Meets Religion, Nov. 29, 2022; last updated Sept. 24, 2024. https://www.sciencemeetsreligion.org/2022/11/do-probability-arguments-refute-evolution-2/.
Basic Biology. "Taxonomy." Basic Biology, n.d. https://basicbiology.net/biology-101/taxonomy.
Betts, Jonathan D., et al. "Second." *Britannica*, July 20, 1998; last updated Nov. 20, 2024. https://www.britannica.com/science/second.
Biological Principles. "Origin of Life on Earth." Biological Principles, n.d. https://bioprinciples.biosci.gatech.edu/module-1-evolution/origin-of-life/.
Bolelli, Daniele. *Create Your Own Religion: A How-To Book Without Instructions*. Disinformation, 2013.
CDC Suicide Prevention. "Facts About Suicide." CDC Suicide Prevention, July 23, 2024. https://www.cdc.gov/suicide/facts/index.html.
Centrone, Bruno. "Pythagoras." Encyclopedia, last updated June 8, 2018. https://www.encyclopedia.com/people/philosophy-and-religion/philosophy-biographies/pythagoras.
Chaitin, Gregory. "The Perfect Language." *Inference* 1 (2015). https://inference-review.com/article/the-perfect-language.
Chambers, Richard L., and Jeffrey M. Yarus. "Geologically Based, Geostatistical Reservoir Modeling." In *Emerging and Peripheral Technologies*, edited by H. R. Warner Jr., 51–111. Vol. 6 of *Petroleum Engineering Handbook*. Society of Petroleum Engineers, 2007.

Bibliography

Cherry, Kendra. "What Is Cross-Cultural Psychology?" Very Well Mind, last updated Dec. 16, 2023. https://www.verywellmind.com/what-is-cross-cultural-psychology-2794903.

Cranston, Maurice, et al. "David Hume." *Britannica*, July 26, 1999; last updated Nov. 20, 2024. https://www.britannica.com/biography/David-Hume.

Difference Between. "Positivism and Post-Positivism." Difference Between, Apr. 23, 2015. https://www.differencebetween.com/difference-between-positivism-and-vs-post-positivism/.

Donne, John. *Devotions upon Emergent Occasions*. Ann Arbor Paperbacks. University of Michigan Press, 2011. Ebook.

Dostoyevsky, Fyodor. *The Brothers Karamazov*. Translated by Constance Garnett. Lowell, 2009. Ebook.

Editors of *Encyclopaedia Britannica*, The. "Fermat's Last Theorem." *Britannica*, July 20, 1998; last updated Dec. 20, 2024. https://www.britannica.com/science/Fermats-last-theorem.

———. "Georg Cantor." *Britannica*, July 20, 1998; last updated Dec. 2, 2024. https://www.britannica.com/biography/Georg-Ferdinand-Ludwig-Philipp-Cantor.

———. "Newton's Laws of Motion." *Britannica*, July 20, 1998; last updated Nov. 5, 2024. https://www.britannica.com/science/Newtons-laws-of-motion.

Energy Education. "Entropy." Energy Education, n.d. https://energyeducation.ca/encyclopedia/Entropy.

Feigl, Herbert, et al. "Positivism." *Britannica*, Sept. 28, 1998; last updated Nov. 19, 2024. https://www.britannica.com/topic/positivism.

Graham, Daniel W. "Heraclitus." *Stanford Encyclopedia of Philosophy*, Feb. 8, 2007; last updated Dec. 8, 2023. Edited by Edward N. Zalta and Uri Nodelman. https://plato.stanford.edu/entries/heraclitus/.

Graziano, Michael. "A New Theory Explains How Consciousness Evolved." *Atlantic*, June 6, 2016. https://www.theatlantic.com/science/archive/2016/06/how-consciousness-evolved/485558/.

Hume, David. *A Treatise of Human Nature*. Edited by L. A. Selby-Bigge. Oxford University Press, 1960.

Huxley, Aldous. *Brave New World*. Huxley, 1932. https://www.huxley.net/bnw.pdf.

Kraut, Richard, et al. "Socrates." *Britannica*, July 26, 1999; last updated Dec. 20, 2024. https://www.britannica.com/biography/Socrates.

Lawler, Katie. "Why Do Birds Pretend to Be Injured?" Birdful, Jan. 31, 2024. https://www.birdful.org/why-do-birds-pretend-to-be-injured/.

Lewis, C. S. *The Problem of Pain*. Macmillan, 1947.

Livio, Mario. *The Golden Ratio: The Story of Phi, the World's Most Astonishing Number*. Broadway, 2002.

Loeb, Abraham, et al. "Relative Likelihood for Life as a Function of Cosmic Time." *Journal of Cosmology and Astroparticle Physics* (2016). https://doi.org/10.1088/1475-7516/2016/08/040.

Bibliography

Mann, David. "Post Positivism." 1Library, 2014. From "A Mixed Methods Evaluation of the Emotional Literacy Support Assistants (ELSA) Project" (DAppEdPsy diss., University of Nottingham), 61–65. https://1library.net/article/post-positivism-epistemological-paradigms.q2pvo6jy.

McCombs, Charles. "Evolution Hopes You Don't Know Chemistry: The Problem with Chirality." Institute for Creation Research, May 1, 2004. https://www.icr.org/article/evolution-hopes-you-dont-know-chemistry-problem-wi/.

Morris, Henry M. "The Mathematical Impossibility of Evolution." Institute for Creation Research, Nov. 1, 2003. https://www.icr.org/article/mathematical-impossibility-evolution/.

Müller, Kasper. "Transcendental Numbers." Cantor's Paradise, Jan. 10, 2021. https://www.cantorsparadise.com/transcendental-numbers-9d4bbe6507cb.

Musielak, Dora E. "Euler: Genius Blind Astronomer Mathematician." Arxiv, n.d. https://arxiv.org/pdf/1406.7397.

NASA. "Goldilocks Zone." NASA, Oct. 23, 2015. https://science.nasa.gov/resource/goldilocks-zone/?ref=the-e-world.

National Park Service. "Martin Luther King, Jr. Memorial: District of Columbia." National Park Service, n.d. https://www.nps.gov/mlkm/learn/quotations.htm.

O'Connor, J. J., and E. F. Robertson. "A History of Zero." MacTutor, Nov. 2000. https://mathshistory.st-andrews.ac.uk/HistTopics/Zero/.

PHILO-Notes. "Aristotle's Concept of the Self." PHILO-Notes, May 16, 2022. https://philonotes.com/tag/aristotles-concept-of-the-self.

Piccirillo, Ryan A. "The Lockean Memory Theory of Personal Identity: Definition, Objection, Response." *Inquiries* 2 (2010). http://www.inquiriesjournal.com/articles/1683/the-lockean-memory-theory-of-personal-identity-definition-objection-response.

Robinson, Lizzie. "What Is Pi?" Fact Site, Jan. 27, 2020; Apr. 30, 2021. https://www.thefactsite.com/what-is-pi/.

Rothman, Lily. "Here's What Beethoven Did When He Lost His Hearing." *Time*, Dec. 16, 2015. https://time.com/4152023/beethoven-birthday/.

Shakespeare, William. *As You Like It*. Global Grey, 2018. Ebook.

———. *Romeo and Juliet*. Edited by Barbara A. Mohat and Paul Werstine. Folger Shakespeare Library, Dec. 2, 2022. https://shakespeare.folger.edu/shakespeares-works/romeo-and-juliet/.

Shuttleworth, Martyn. "Hawthorne Effect." Explorable, Oct. 10, 2009. https://explorable.com/hawthorne-effect.

Simon, Paul. "I Am a Rock." Track 1 on *The Paul Simon Songbook*. Recorded Aug. 1965. LP.

Stackhouse, John G., Jr. *Can God Be Trusted? Faith and the Challenge of Evil*. Oxford University Press, 2000.

Bibliography

Szalay, Jessie. "Perpetual Motion Machines: Working Against Physical Laws." Live Science, Aug. 30, 2016. https://www.livescience.com/55944-perpetual-motion-machines.html.

Trout Fishing in America. "Back When I Could Fly." Track 14 on *Family Music Party*. Trout Records 1998. Compact disc.

USA. "Official Language of the United States." USA, last updated Mar. 3, 2025. https://www.usa.gov/official-language-of-us.

Watson, Richard A., et al. "René Descartes." *Britannica*, Aug. 5, 1998; last updated Dec. 9, 2024. https://www.britannica.com/biography/Rene-Descartes.

Weaver, Richard M. *Ideas Have Consequences*. University of Chicago Press, 1948.

Weizmann Institute of Science. "Quantum Theory Demonstrated: Observation Affects Reality." Science Daily, Feb. 27, 1998. https://www.sciencedaily.com/releases/1998/02/980227055013.htm.

West, Catherine. "How Culture Affects the Way We Think." Association for Psychological Science, Aug. 1, 2007. https://www.psychologicalscience.org/observer/how-culture-affects-the-way-we-think.

Ye, Xiaojing, and The Conversation. "The Long Search for the Value of Pi." *Scientific American*, Mar. 14, 2016. https://www.scientificamerican.com/article/the-long-search-for-the-value-of-pi/.

Zoltán, Imre, et al. "Ignaz Semmelweis." *Britannica*, July 20, 1998; last updated Nov. 28, 2024. https://www.britannica.com/biography/Ignaz-Semmelweis.

Scripture Index

Genesis

1:1	51
1:3	51
1:6	51
1:9	51
1:11	51
1:14	51
1:20	51
1:24	52
1:26	53
1:26–31	75
2:7	53
2:16–17	54
2:18	75
3	71
3:1–19	54
3:4–5	71
3:6	71
5	44
11	71
11:4	71
12:3	72, 79

Exodus

3:11	72
3:13	52–53
3:14	53

Leviticus

11:44	74
17:11	58

Judges

6:15	72

1 Samuel

7	45
7:12	45
16:1–4	72
16:5–6	72
16:7	72
16:8–10	72
16:11	72
16:12	72–73
17:33	73
17:34–37	73

2 Samuel

7:16	79
24	61
24:24	61

1 Kings

7:23	76

Scripture Index

1 Chronicles
29:10–20 — v

Esther
4:14 — 73

Job
38:1–3 — 109
38:4–7 — 109

Psalm
8:3–4 — 108
19:1–4 — 50
34:8 — 93
107:13–14 — 97
118:22 — 119

Proverbs
1:1–7 — 77
14:8 — 57
14:12 — 48, 57
16:18 — 57
16:25 — 57
26:12 — 57
27:17 — 124

Ecclesiastes
1:2 — 127
3:11 — 72
4:12 — 124
12:12–14 — 127
12:13–14 — 40

Isaiah
6:1 — 108
6:5 — 108
6:8 — 104
6:9–10 — 78
28:16 — 119
59:9–18 — 99–100
64:6 — 59

Jeremiah
31:33 — 123

Ezekiel
3:4–6 — 102–103
3:21 — 103
4 — 103
5 — 103

Daniel
2:20–23 — 77–78
2:26–28 — 78

Hosea
3 — 104

Amos
7:7–8 — 57

Habakkuk
2:4 — 47

Matthew
5:48 — 74
10:16 — 66
11:28–30 — 109–110
13:11–15 — 78
16 — 110
16:13–20 — 118–119
18:1–4 — 115
18:2–4 — 65
18:12–14 — 108
16:24–26 — 110
20:16 — 111
26:36–46 — 45

Mark
9:23–24 — 47
10:35–45 — 115
12:28–31 — 74

Scripture Index

Luke

6:39	7
9:22	67
12:19	38
12:20	38
16	46
16:10–12	73
19:1–9	110

John

1:1–4	52
1:14	52, 102
1:16–17	52
3	66
3:3	84
3:16	53
3:17	90
3:19	57
3:26–30	110
4:13–14	58
5:24	84
8	89
8:10–11	90
8:11	126
8:12	97
10	108
11	46, 67
11:25	97
11:39	85
11:39–41	85
11:41–42	85
11:43	85
11:44	85
18:37	97
19:30	125
20	47
20:18	67
20:25	67
20:27	67–68
20:29	67
20:30–31	66–67
21:25	66

Acts

4:11	119
17	97
17:6	97
17:10–15	69
26	125
26:28–29	125

Romans

1:22	7
3–6	66
3:3–4	52
3:23	57
5:6–11	59, 126
6:5	126
6:23	57
7:15–25	57, 94
7:24–25	94
8:29	119
10:9–10	59, 61
10:13–17	104
12:1–2	99
12:3–8	124

1 Corinthians

1:18	128
1:26–31	109
3	119
3:1–6	119–120
3:18	66
4:1–2	79
8:8–13	104
9:19–24	104–105
11:1	v
12	124
13:11	65, 81
13:12	69
13:1–13	106
15:19	126, 129

2 Corinthians

4:1–6	97–98

Scripture Index

2 Corinthians (continued)

5:17	89
10:5	111
11:3	66
12:9	109

Galatians

1:6–7	66
5:19–21	91
5:22–23	91
5:24–26	91

Ephesians

1:7–9	108
2	66
2:6–7	75, 91
2:6–8	125
2:8–9	59, 122
2:10	75, 91
2:19–22	122
2:20	119
3:4	78
3:5	79
3:6–7	79
3:8–9	79
3:10–11	79–80
3:14–19	81
3:20–21	81
4:1	74
4:1–3	92
4:17–19	85
4:20–24	85–86
4:25–32	86
5:6–7	66
6:12	98

Philippians

2:1–11	98
2:5–8	112
2:12	69, 81
2:12–13	65
2:13–16	98

3:7–9	115
3:12	94, 128
3:13–14	128
3:15	128
3:17	129
3:18–19	128
4:11–12	92
4:13	92

Colossians

1:24–28	80
2:2–3	82
2:4	66

1 Timothy

3:16	80
4:7–8	87, 92

2 Timothy

2:15	81

Hebrews

5:12–14	86
5:13–14	90
8:10	123
9–10	66
9:22	58
10:16	123
10:23–25	124
11	45
11:1–2	43
11:3	43, 50
11:4	44
11:6	44
11:13	45
11:38	44
12:1	45
12:2	46
12:1–2	92

Scripture Index

James
1:14–15	57
1:27	123
2:18	130
2:26	126

1 Peter
1:18–19	59
1:14–15	74
2:1–3	92–93
2:3	94
2:4–5	93
2:6–7	119
2:9	93
2:11–12	93
2:21–25	93

2 Peter
1:3–4	87
1:5–8	87

1 John
3:1	108
4:14	97

www.ingramcontent.com/pod-product-compliance
Lightning Source LLC
Chambersburg PA
CBHW072147160426
43197CB00012B/2284